HYDRAULICS

HYDRAULICS

LABORATORY MANUAL

S.K. LIKHI

Technical Teachers Training Institute
Chandigrah

PUBLISHING FOR ONE WORLD

NEW AGE INTERNATIONAL (P) LIMITED, PUBLISHERS

New Delhi • Bangalore • Chennai • Cochin • Guwahati • Hyderabad
Jalandhar • Kolkata • Lucknow • Mumbai • Ranchi

Visit us at **www.newagepublishers.com**

Published by New Age International (P) Ltd., Publishers
First Edition: 1995
Reprint: 2012

BRANCHES

- **Bangalore** 37/10, 8th Cross (Near Hanuman Temple), Azad Nagar, Chamarajpet, Bangalore-560 018
  Tel.: (080) 26756823, Telefax: 26756820, E-mail: bangalore@newagepublishers.com
- **Chennai** 26, Damodaran Street, T. Nagar, Chennai-600 017
  Tel.: (044) 24353401, Telefax: 24351463, E-mail: chennai@newagepublishers.com
- **Cochin** CC-39/1016, Carrier Station Road, Ernakulam South, Cochin-682 016
  Tel.: (0484) 2377004, Telefax: 4051303, E-mail: cochin@newagepublishers.com
- **Guwahati** Hemsen Complex, Mohd. Shah Road, Paltan Bazar, Near Starline Hotel, Guwahati-781 008
  Tel.: (0361) 2513881, Telefax: 2543669, E-mail: guwahati@newagepublishers.com
- **Hyderabad** 105, 1st Floor, Madhiray Kaveri Tower, 3-2-19, Azam Jahi Road, Nimboliadda, Hyderabad-500 027
  Tel.: (040) 24652456, Telefax: 24652457, E-mail: hyderabad@newagepublishers.com
- **Kolkata** RDB Chambers (Formerly Lotus Cinema) 106A, 1st Floor, S.N. Banerjee Road, Kolkata-700 014
  Tel.: (033) 22273773, Telefax: 22275247, E-mail: kolkata@newagepublishers.com
- **Lucknow** 16-A, Jopling Road, Lucknow-226 001
  Tel.: (0522) 2209578, 4045297, Telefax: 2204098, E-mail: lucknow@newagepublishers.com
- **Mumbai** 142C, Victor House, Ground Floor, N.M. Joshi Marg, Lower Parel, Mumbai-400 013
  Tel.: (022) 24927869, Telefax: 24915415, E-mail: mumbai@newagepublishers.com
- **New Delhi** 22, Golden House, Daryaganj, New Delhi-110 002
  Tel.: (011) 23262368, 23262370, Telefax: 43551305, E-mail: sales@newagepublishers.com

ISBN (10) : 81-224-0516-9
ISBN (13) : 978-81-224-0516-3

₹ 100.00

C-12-06-6376

Printed in India at Saras Graphics, Rai, Haryana.

PUBLISHING FOR ONE WORLD

NEW AGE INTERNATIONAL (P) LIMITED, PUBLISHERS
7/30 A, Darya Ganj, New Delhi-110002
Visit us at **www.newagepublishers.com**

Foreword

Development of capabilities of related practical skills among technician engineers is very vital for their effective performance in the world of work. As such, the learning of practical skills need to be considered as an integral part of the instructional process for any technician engineering programme.

The institute has for the many years been engaged in the development of laboratory manuals for introducing teaching-learning practices in the polytechnics which would help achieve the above objective of training technician engineers. This laboratory manual on Hydraulics is one of the attempts in this direction. Professor S.K. Likhi, the author of this laboratory manual has worked extensively in this area and tried out this manual in the polytechnics before presenting it in the present form. I hope that this laboratory manual will prove useful to both the teachers and engineering students in the teaching and learning of Hydraulics.

M.M. MALHOTRA
PRINCIPAL

Preface

The experiments described in this manual are part of the curriculum of hydraulics for the three-year diploma course in Civil Engineering and of Fluid Mechanics-I for degree courses in various branches of engineering. It explains the procedures of performing various experiments in the hydraulics laboratory. The use of this manual in conducting the laboratory work described will lead the students to the preparation of a laboratory report which is sufficiently detailed and exhaustive. It is a convenient way of imparting instruction for handling apparatus, indicating the range and accuracy of observations, and providing a guide to the presentation of results. The students are required to interpret the result of the experiment, with a view to make them appreciate the importance and significance of the test in real-life situations.

The manual is complete in itself and the required graph sheets, etc. form an integral part of each experiment included. For each experiment, the observations can be entered on the tabulated sheets, calculations made and the results plotted on the graph sheets attached. Questions given at the end of experiments can lead to better comprehension if these are discussed by the teacher with the students. It is expected that this manual will help both teachers and students effectively understand the principles of hydraulics. Despite our best efforts, some errors may still remain. I shall be grateful if these are brought to my notice. Suggestions for improvement are welcome.

LIKHI, S.K.

Acknowledgements

In bringing out the manual, in this form, I gratefully acknowledge the continuous guidance and inspiration of Prof. R. Subramanian, Head of Civil Engineering Department. Thanks are also due to polytechnic teachers for using this manual in its developmental stage and for their comments for improvement, and to Shri Pritam Singh and Shri Kishan Singh for drafting and typing, respectively.

I am also grateful to the Indian Standards Institution for granting us permission to refer to Indian Standards.

LIKHI, S.K.

Contents

Measurement Techniques

INTRODUCTION

Before one starts working in the hydraulics laboratory, one must have the knowledge of physical properties of fluids and the behaviour of fluid at rest or in motion and have the skill of measurement of pressure and of other variable factors such as temperature, depth of liquid level, etc. Various fluid properties required to be known are given in Appendix I. Most of the measurement skills such as measurement of length, breadth, angle, temperature, etc., are very elementary and are learnt by the students in their school days. Measurements of pressure at a point, difference of pressure between two points and discharge measurements are some of the skills which a student is supposed *to know but is not given the time to learn;* though theoretical aspect of these skills is taught by the teacher in his lecture class.

Fluid pressure at any point is given by the formula:

$$p = w\,h$$

where p = fluid pressure at a point

w = specific weight of fluid

and h = depth of the point from free liquid surface, also called pressure head

Thus pressure head or head of liquid, h, is given by $h = \frac{p}{w}$

In hydraulics, pressure is measured either by manometers or by gauges. Bourden gauge is the most commonly used gauge for measurement of fluid pressure whereas the following manometers are commonly used in hydraulics laboratory for measurement of pressure at a point or pressure difference between any two points.

(i) Piezometric tube
(ii) Double-column manometer
(iii) Single-tube manometer
(iv) Differential manometer
(v) Inverted u-tube manometer
(vi) Bourdon tube pressure gauge

Let us go to the hydraulics laboratory and study the aforementioned manometers and pressure gauges for learning the skill in measurement of

(i) Pressure at a point
(ii) Pressure difference between any two points, and
(iii) Discharge

PRESSURE MEASUREMENT AT A POINT

Piezometric Tube: Simple piezometric tubes are used for measurement of pressure at a point. A piezometric tube is tapped into the wall of the container or conduit as shown in Fig. 0.1. The rise of the liquid column, h, in the piezometric tube is measured and the pressure p at the gauge point A is given by $p = wh$ where w is the specific weight of the liquid. Value of h is measured with the help of meter rod attached along the piezometric tube.

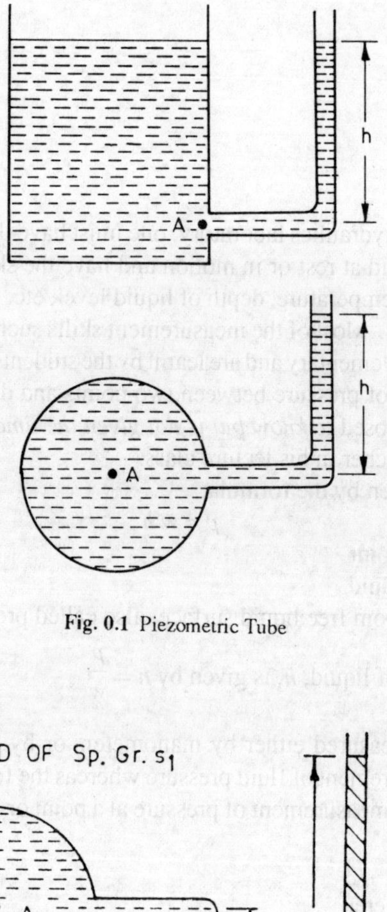

Fig. 0.1 Piezometric Tube

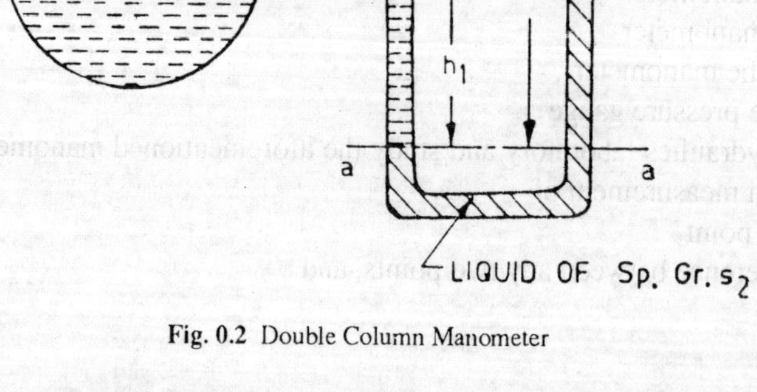

Fig. 0.2 Double Column Manometer

Double Column Manometer: This type of manometer consists of single U-tube, one end of which is connected to the gauge point and the other end is open to the atmosphere as shown in Fig. 0.2. Water, mercury, oil, salt solution, carbon-disulphide, bromoform, alcohol, carbon tetra-chloride are the commonly used measuring liquids.

If s_1 is the specific gravity of fluid flowing in the pipe and s_2 is the specific gravity of measuring liquid, then the pressure at point A is given by:

$$p_A + ws_1h_1 = ws_2h_2$$

$$\therefore \quad p_A = ws_2h_2 - ws_1h_1$$

$$= w(s_2h_2 - s_1h_1)$$

where h_1 and h_2 are as shown in the figure and are measured with the help of meter rods attached along glass tubes.

Single Tube Manometer: It is a modified form of double-column manometer in which cross-sectional area of one limb is so large as compared to that of the other limb that any change in level of the former may be neglected.

The manometer is connected to the gauge point and the change in level of the liquid in the graduated small diameter limb is noted; neglecting the change in the level of liquid in the large cross-sectional area limb called reservoir or basin.

The graduated leg of this manometer may be vertical or inclined depending upon the accuracy required. Accordingly we have (a) manometer with vertical leg and (b) manometer with inclined leg.

(a) *Manometer with Vertical Leg:* The pressure at the gauge point A, as shown in Fig. 0.3, is given by

$$p_A + s_1w(y + \delta y) = ws_2(h + \delta y)$$

$$p_A = ws_2(h + \delta y) - ws_1(y + \delta y)$$

Since

$$\delta y = \left(\frac{a}{A}\right)h$$

i.e.,

$$\delta y = \left(\frac{\text{area of tube}}{\text{area of reservoir}}\right)h \text{ we get}$$

$$p_A = ws_2\left(h + \frac{a}{A}h\right) - ws_1\left(y + \frac{a}{A}h\right)$$

Since $\frac{ah}{A}$ is negligibly small, we get

$$p_A = w(s_2h - s_1y)$$

(b) *Manometer with Inclined Leg:* If θ is the inclination of the inclined tube with the horizontal as shown in Fig. 0.4, the pressure at point A is given by

$$p_A + ws_1(y + \delta y) = ws_2(h \sin\theta + \delta y)$$

$$p_A = ws_2(h \sin\theta + \delta y) - ws_1(y + \delta y)$$

$$= ws_2\left(h\sin\theta + \frac{a}{A}h\right) - ws_1\left(y + \frac{a}{A}h\right)$$

If the ratio $\frac{a}{A}$ is negligibly small, the pressure at point A is given by

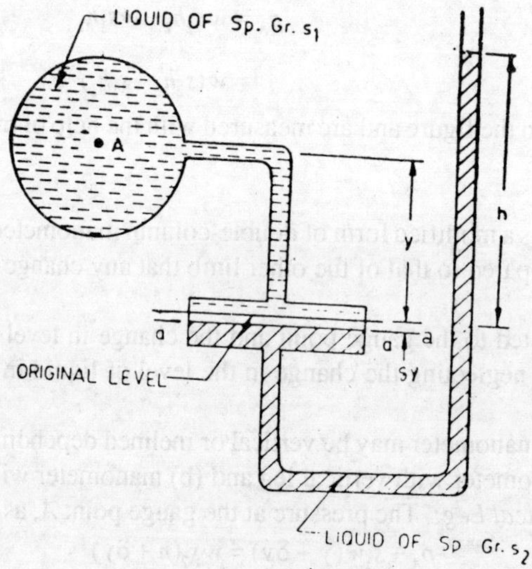

Fig. 0.3 Single Column Manometer

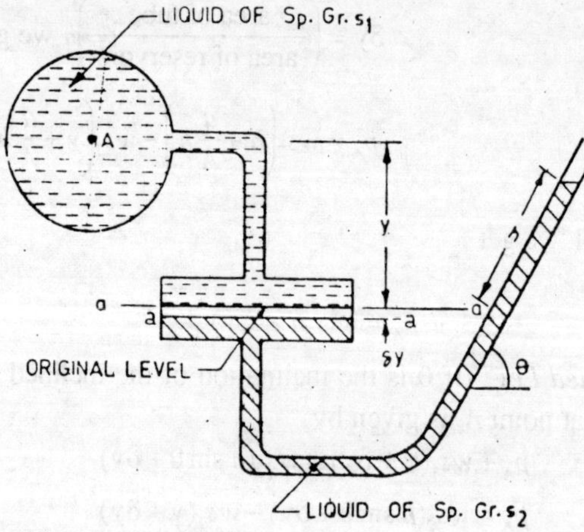

$$p_A = ws_2h\sin\theta - ws_1y$$

$$= w(s_2h\sin\theta - s_1y)$$

MEASUREMENT OF PRESSURE DIFFERENCE BETWEEN TWO POINTS

Differential Manometer: If p_A and p_B are the pressure intensities at the centres of pipes A and B respectively as shown in Fig. 0.5, then

$$ws_1h_1 + p_A = ws_2h_2 + ws_3h_3 + p_B$$

$$p_A - p_B = ws_2h_2 + ws_3h_3 - ws_1h_1$$

$$p_A - p_B = w(s_2h_2 + s_3h_3 - s_1h_1)$$

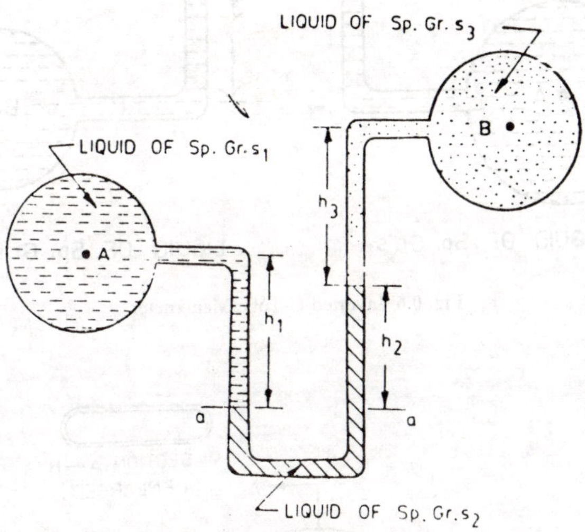

Fig. 0.5 Differential Manometer

where s_1 and s_3 are specific gravities of liquids flowing in pipes A and B respectively and s_2 is specific gravity of measuring liquid, h_1, h_2 and h_3 are measured with the help of meter rods attached along the glass tubes.

Inverted U-tube Manometer: This type of manometer consists of an inverted U-tube connected to the gauge points of the same pipeline or of two different pipes. An opening at the top of the inverted U-tube is provided for driving out any air present in the pipeline. The tube of manometer may also be filled with the liquid lighter than the liquid flowing in the pipelines as shown in Fig. 0.6. The pressure difference between gauge points A and B is given by:

$$p_A - ws_1(y + h) = p_B - ws_2y$$

$$= ws_1 y + ws_1 h - ws_2 y$$

$$= w(s_1 - s_2)y + ws_1 h$$

(pressure due to air, being negligible, is omitted)

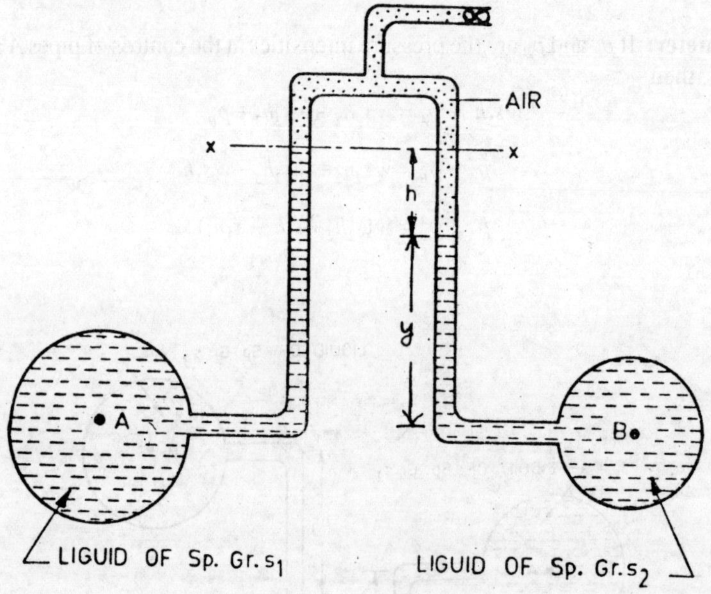

Fig. 0.6 Inverted U-Tube Manometer

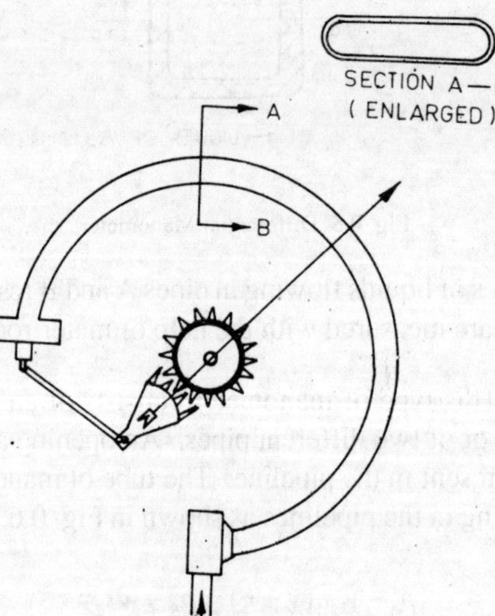

Bourdon Tube Pressure Gauge: This type of gauge is used to measure pressure at a point. Naturally two such gauges will be used for measuring pressure difference between any two gauge points. The pressure gauge is screwed to the gauge point. The fluid under pressure enters the curved tube at the fixed end and tends to straighten out this tube. The movement of the free end of the tube is transferred to the needle by means of levers, rack and pinion arrangements, and the needle moves to indicate the pressure on a properly calibrated scale as shown in Fig. 0.7.

Observations and Calculations

Specific gravity of manometer liquid = s_2
Specific gravity of liquid flowing in pipe = s_1

S.No.	Type of manometer/gauge	Manometer reading			Formula	Pressure/pressure difference
		h_1	h_2	$h = h_2 - h_1$		
	Units					
1.	Piezometric tube					
2.	Double-column manometer					
3.	Single-tube manometer (a) Vertical tube (b) Inclined tube					
4.	Differential manometer					
5.	Inverted U-tube manometer					
6.	Bourden gauge					

MEASUREMENT OF WATER SURFACE LEVEL

Position of water surface level is measured with the help of pointer gauge which is a circular rod having one end attached to the scale and the other being a pointed one. The pointed end is made to touch the water surface and the reading is taken from the attached scale. The rise or fall in water-level is equal to the final water surface reading minus initial water surface reading. Rise or fall in water-level multiplied by area of the collecting tank will give the volume of water collected or discharged.

METHOD TO FIND THE LEAST COUNT OF HOOK GAUGE

Let the zero of vernier scale coincide with the zero of the main scale and we find the number of division of vernier scale which coincides with a division of main scale. Say the 10th division of vernier scale coincides with the 9th division of the main scale, then

10 **VSD** (vernier scale division) = 9 **MSD** (main scale division)

$$1\ \text{VSD} = \frac{9}{10}\ \text{MSD}$$

Since least count of vernier scale is defined as difference of 1 **MSD** and 1 **VSD.**

Least count = 1 **MSD** − 1 **VSD**

$$\text{Least count} = 1\ \text{MSD} - \frac{9}{10}\ \text{MSD}$$

$$= \frac{1}{10}\ \text{MSD}$$

$$= 0.1\ \text{mm of MSD}$$

Hence, least count of vernier = 0.1 mm

= 0.01 cm

In case zero of vernier scale lies between 7.4 and 7.5 cm and 6th division of vernier scale coincides with main scale, then the hook gauge reading will be $7.4 + 6 \times 0.01 = 7.46$ cm.

MEASUREMENT OF DISCHARGE

Quantity of fluid flowing per unit time is known as discharge and is measured either by weight or by volume. Volumetric measurement of discharge is the most commonly used method in hydraulics laboratories.

Discharge Measurement by Weight: Flow from the apparatus is collected in a weighing tank. A known weight of liquid is collected in the weighing tank and the duration of flow is noted with the help of a stop-watch.

Observations and Calculations

Weight of water collected, W =

Duration of collection, t =

Discharge, $Q = \frac{W}{t}$ kg/s =

Discharge Measurement by Volume: Flow from the apparatus is collected in a collecting tank. Rise of liquid level in the tank is measured either by a graduated piezometric tube fitted to one of the tank walls or by means of a pointer gauge.

Observations and Calculations

Say surface area of tank, A =

Initial water-level reading, h_1 =

Final water-level reading, h_2 =

Rise in water-level, $(h_2 - h_1)$ =

Volume of water collected, $A\,(h_2 - h_1)$ =

Duration of collection, t =

Discharge, $Q = A\,\frac{(h_2 - h_1)}{t}$ =

Precautions

(i) Since the water in the glass tube has concave upward meniscus and mercury in glass tube has convex upward meniscus, mamometer readings should be taken carefully.

(ii) Care should be taken to remove parallax before taking reading so that the pointed end of the hook gauge just touches the pointed end of its reflection in water.

Floating Bodies

EXPERIMENT 1

To determine the metacentric height of a given vessel under:

(a) Unloaded condition, and

(b) Loaded condition

Theory: The stability of a body such as a ship which floats in the surface of a liquid, is of great importance. Whether the equilibrium is stable, neutral or unstable is determined by the height of its centre of gravity. Consider the rectangular pontoon shown floating in equilibrium on even keel as shown in cross-section in Fig. 1.1 (a).

The weight of the pontoon acts vertically downwards through its centre of gravity, G, and this is balanced by an equal and opposite buoyant force acting upwards through the centre of buoyancy, B, which lies at the centre of gravity of liquid displaced by the pontoon.

Corresponding to a small angular displacement $\delta\theta$ from the equilibrium position as shown in Fig. 1.1(b) due to displacement δx of jockey weight, the centre of gravity of the liquid displaced by the pontoon shifts from B to B_1 and the vertical line of action of the buoyant force intersects the extension of the line BG in M, the metacentre.

The equal and opposite forces through G and B_1 exert a couple on the pontoon, and provided that M lies above G as shown in Fig. 1.1(b), *this couple acts in the sense of restoring the pontoon to even keel, i.e, the pontoon is stable.* If, however, the matacentre M lies below the centre of gravity G, the sense of the couple is to increase the angular displacement and the pontoon is unstable. The special case of neutral stability occurs when M and G coincide.

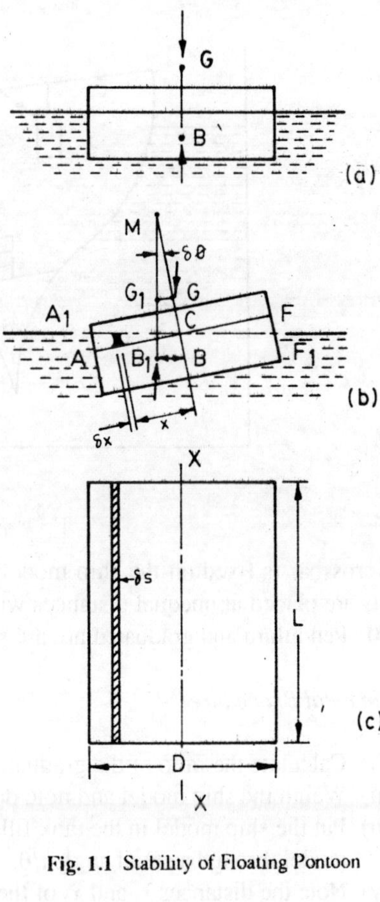

Fig. 1.1 Stability of Floating Pontoon

The distance GM is given by

$$GM = \frac{w}{W}\frac{\delta x}{\delta\theta}$$

where w is the jockey weight displaced by distance δx and W is the weight of the pontoon, or in the limit,

$$GM = \frac{w}{W}\frac{dx}{d\theta}$$

Apparatus: Figure 1.2 shows the arrangement in which a ship model is made to float in a small steel tank 400 mm × 400 mm × 400 mm, filled with water and having at least one side wall made of glass. Removable steel strips are placed in the ship model for the purpose of changing its weight. By means of a pendulum and the graduated arc, the angle of tilt, θ , can be measured.

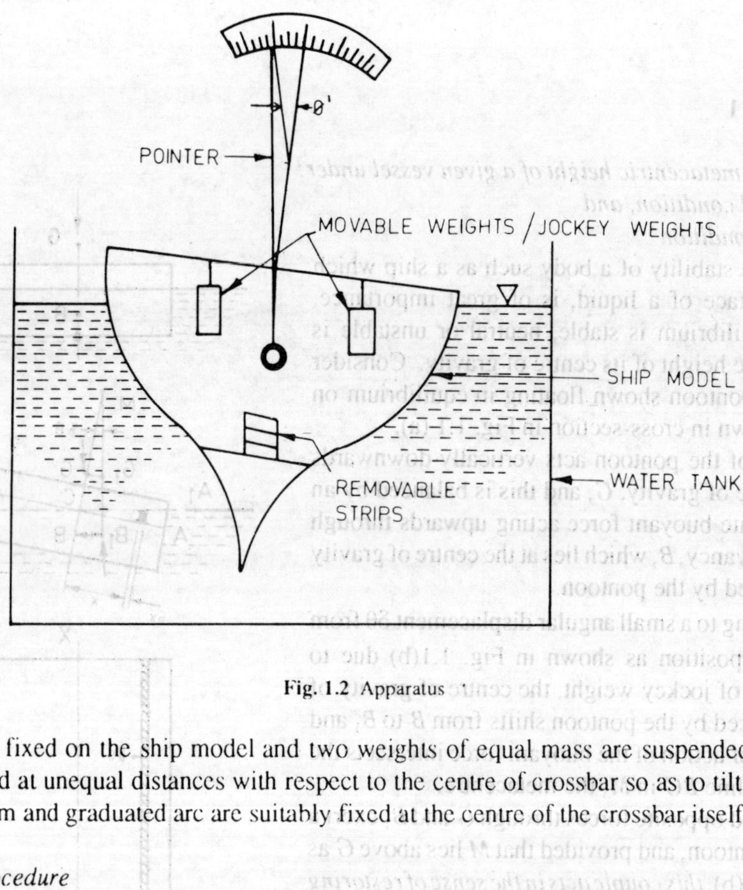

Fig. 1.2 Apparatus

A crossbar is fixed on the ship model and two weights of equal mass are suspended from it. These weights are placed at unequal distances with respect to the centre of crossbar so as to tilt the model by an angle θ. Pendulum and graduated arc are suitably fixed at the centre of the crossbar itself.

Experimental Procedure

(i) Calculate the ratio of the graduated arc to the length of the pointer, say n.
(ii) Weigh the ship model and note down its value, say W.
(iii) Put the ship model in the tank filled with water and displace the movable massess so as to tilt the model through a small angle $d\theta$.
(iv) Note the distances x_1 and x_2 of the masses from the centre of crossbar and find dx.

(v) Read the angle θ' on the graduated scale and calculate $d\theta$ where $d\theta = n\theta'$.

(vi) Repeat steps (iii), (iv) and (v) for different positions of movable masses.

(vii) Repeat steps (ii) to (vi) for three different weights of the ship model. (Weight of the ship model is varied by removing the steel strips.)

Observations and Calculations

Ratio of the radius of the graduated arc to the length of the pointer $= n$

Weight of the movable mass, w =

S.No.		Weight of ship model, W	x_1	x_2	$dx = x_1 - x_2$	θ'	$d\theta = n\theta'$	$GM = \frac{wdx}{Wd\theta}$	Average metacentric height
	Units								
1.	Unloaded condition								
2.	Loaded condition								
3.	Loaded condition								

Discussion of Results

Questions for Discussion

(i) Can you compare the experimental results with those obtained analytically. If yes, list the formula used.

(ii) How would the stability of the pontoon be affected if it were floated on a liquid with a density greater than that of water?

Precautions

(i) Measure the angle of tilt accurately and convert it into radians for calculation purposes.

Fundamentals of Fluid Flow

EXPERIMENT 2

To verify Bernoulli's Theorem

Theory: Daniel Bernoulli enunciated in 1738 that "in any stream flowing steadily without friction, the total energy contained in a given mass is same at every point in its path of flow". This statement is called Bernoulli's theorem.

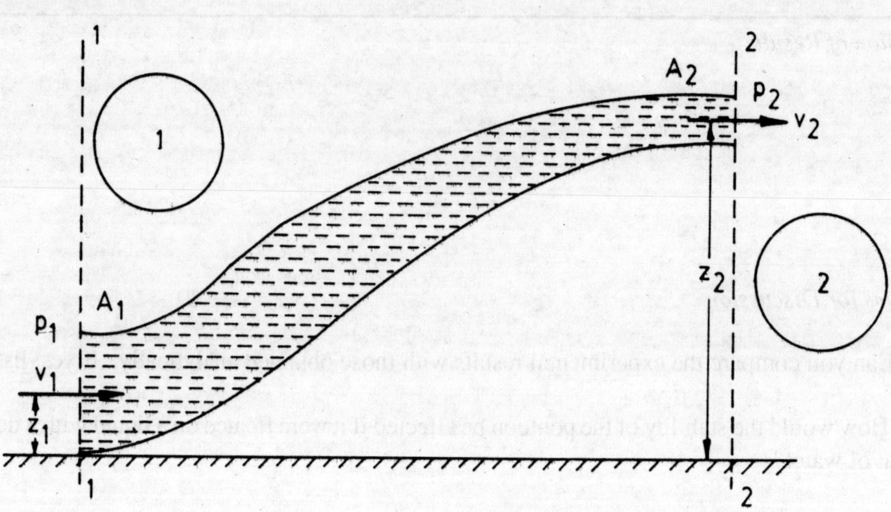

Fig. 2.1 Flow through Stream Tube

With reference to sections 1–1 and 2–2 along the length of steady flow in the stream tube shown in Fig. 2.1, the total energy at section 1–1 is equal to the total energy at section 2–2 as stated in Bernoulli's theorem.

With usual notations, the expression for total energy contained in a unit weight of fluid at sections 1–1 and 2–2 is given by:

$$\text{Total energy at section } 1-1 = \frac{P_1}{w} + \frac{V_1^2}{2g} + Z_1$$

$$\text{Total energy at section } 2-2 = \frac{P_2}{w} + \frac{V_2^2}{2g} + Z_2$$

where $\frac{P_1}{w}$ = pressure energy at section 1 – 1

$\frac{V_1^2}{2g}$ = kinetic energy at section 1 – 1

Z_1 = potential energy at section 1 – 1

and $\frac{P_2}{w}$ = pressure energy at section 2 – 2

$\frac{V_2^2}{2g}$ = kinetic energy at section 2 – 2

Z_2 = potential energy at section 2 – 2

Thus applying Bernoulli's theorem between sections 1 – 1 and 2 – 2 we find:

$$\frac{P_1}{w}+\frac{V_1^2}{2g}+Z_1=\frac{P_2}{w}+\frac{V_2^2}{2g}+Z_2 \qquad (i)$$

In **MKSA** system, the pressure energy, kinetic energy and potential energy are measured in metre of fluid column per unit weight of fluid. Eq. (i) is modified for taking into account the loss of energy due to friction between section 1–1 and 2–2 and is written as

$$\frac{P_1}{w}+\frac{V_1^2}{2g}+Z_1=\frac{P_2}{w}+\frac{V_2^2}{2g}+Z_2+(\Delta H)_{1,2}$$

where $(\Delta H)_{1,2}$ represents the loss of energy between sections 1–1 and 2–2.

Apparatus: It consists of conduit of varying cross-sectional area as shown in Fig. 2.2. It is made of perspex or metal sheets which are jointed together to form a converging rectangular conduit. It is generally 50 mm × 50 mm on the u/s side and 20 mm × 20 mm on the d/s side. The length of the conduit may be 0.5 m. Piezometric tubes are fitted at regular intervals to measure the piezometric head at each gauge point.

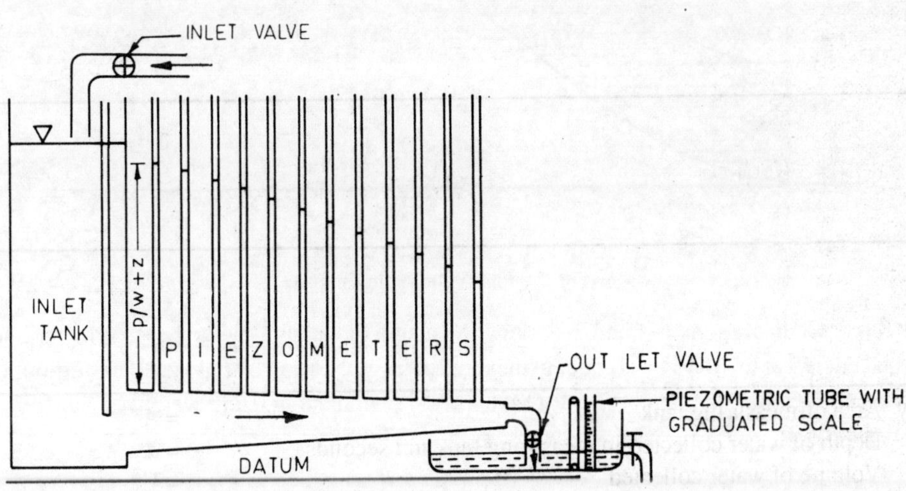

Fig. 2.2 Experimental Set Up for Verification of Bernoulli's Equation

This conduit is connected to an inlet tank. A piezometric tube is also connected to the tank for recording the water-level in tank. An outlet valve is fitted at the downstream end of the pipe. Supply to inlet tank is controlled with an inlet valve. A graph paper is fixed on a vertical board so that the piezometric readings can be taken. Alternatively, a graduated scale may be used for the purpose

Experimental Procedure

(i) Note down the area of the conduit at various gauge points.

(ii) Open the supply valve and adjust the flow so that the water-level in the inlet tank remains constant.

(iii) Measure the height of water-level (above an arbitrarily selected suitable plane) in different piezometric tubes.

(iv) Measure the discharge of the conduit with the help of measuring tank.

(v) Repeat steps (ii) to (iv) for two more discharges.

(vi) Plot graph between total energy and distance of gauge points starting from u/s side of conduit.

Observations and Calculations

(i) Discharge:

Run No.	d	A	$V_0 = Ad$	t	Discharge, Q
Units ⟶					
1					
2					
3					

Say Area of measuring tank $= A$

Depth of water collected in measuring tank in t seconds $= d$

Volume of water collected $= V_0$

(ii) *Total Head*:

	Piezometric tube No.	1	2	3	4	5	6	7	8	9	10
	Distance of gauge point from u/s of conduit										
	Area of cross-section of conduit at each gauge point, A										
Run No. 3, Q =	$V = \frac{Q}{A}$										
	$\frac{V^2}{2g}$										
	$\frac{P}{W} + Z$										
	$\frac{P}{W} + Z + \frac{V^2}{2g}$										
Run No. 2, Q =	$V = \frac{Q}{A}$										
	$\frac{V^2}{2g}$										
	$\frac{P}{W} + Z$										
	$\frac{P}{W} + Z + \frac{V^2}{2g}$										
Run No. 1, Q =	$V = \frac{Q}{A}$										
	$\frac{V^2}{2g}$										
	$\frac{P}{W} + Z$										
	$\frac{P}{W} + Z + \frac{V^2}{2g}$										

Discussion of Results

Questions for Discussion

(i) If piezometric tubes are replaced by pressure gauges, then how will you calculate pressure energy and potential energy of flow?

Precautions

(i) Before taking reading, do take care that steady flow conditions are established in the conduit.
(ii) Check that air is not entrapped in the piezometric tubes.
(iii) There should be no leakage between u/s and d/s end of the conduit.

EXPERIMENT 3

(a) To find the value of the coefficient of a given venturimeter fitted in a pipe and also the discharge of the pipe.

(b) To find the value of the coefficient of discharge for a given orifice-meter and, also, the discharge through a given pipe by the use of orifice-meter.

(a) Venturimeter

Theory: The venturimeter is a device which has been used over many years for measuring the discharge along the pipe. The fluid flowing in the pipe is led through a contracting section to a throat which has a smaller cross-sectional area than the pipe, so that the velocity of fluid through the throat is higher than that in the pipe. This increase of velocity is accompanied by a fall in pressure, the magnitude of which depends on the rate of flow, so that by measuring the pressure drop, the discharge may be calculated. Beyond the throat the fluid is decelerated in a pipe of slowly diverging section, the pressure increasing as the velocity falls.

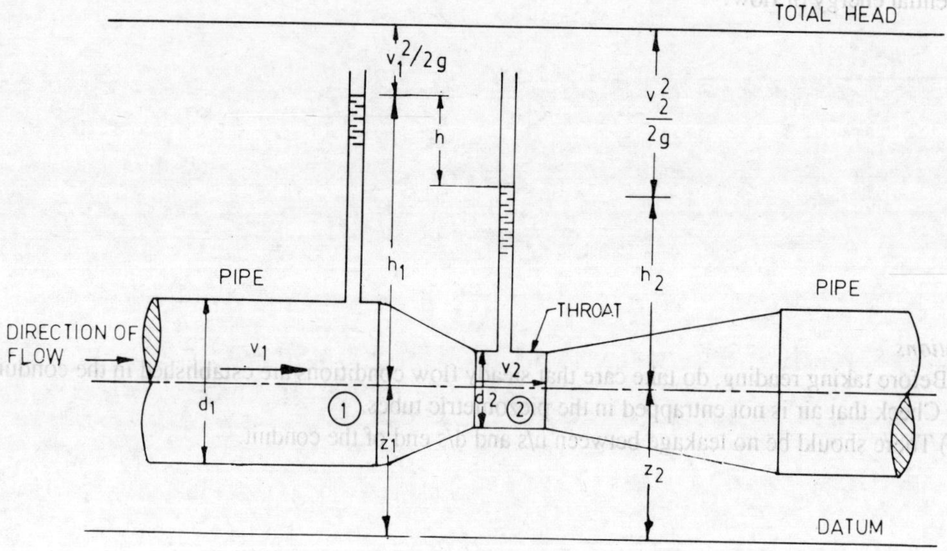

Fig. 3.1 Venturimeter

With usual notations, the expression for discharge, Q, through a venturimeter is given by

$$Q = \frac{a_1 a_2 \sqrt{2g(h_1 - h_2)}}{\sqrt{a_1^2 - a_2^2}}$$

where a_1 = area of cross-section of an upstream section

a_2 = area of cross-section at throat

h_1 = piezometric head at the upstream section

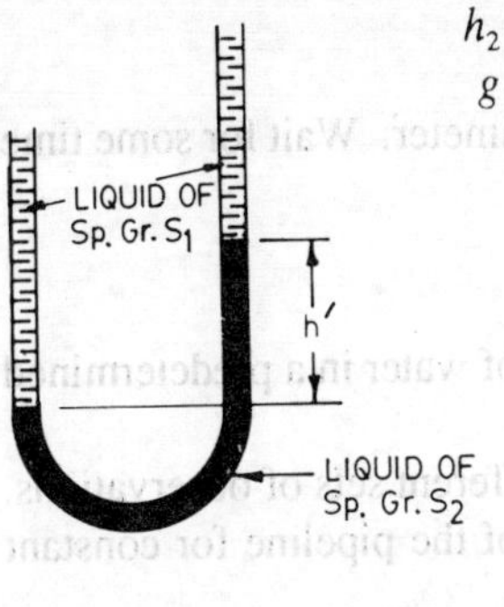

Fig. 3.2 Mercury Manometer

h_2 = piezometric head at the throat as shown in Fig. 3.1

g = acceleration due to gravity

In practice there is some loss of energy between sections 1 and 2 and the velocity is not absolutely constant across either of these sections. As a result, measured value of Q usually falls a little short of those calculated from Eq. (i) and it is customary to allow for this discrepancy by writing:

$$Q = K\left[\frac{a_1 a_2}{\sqrt{a_1^2 - a_2^2}}\right]\sqrt{2g(h_1 - h_2)}$$

$$= \frac{K a_1 a_2}{\sqrt{a_1^2 - a_2^2}}\sqrt{2gh} \text{ where } h = h_1 - h_2 \qquad \text{(ii)}$$

in which K is known as the venturimeter coefficient and its value varies slightly from one venturimeter to another, and even for a given venturimeter it may vary with the discharge, but usually lies within the range of 0.92 to 0.99. In case the piezometric tappings are connected to a differential manometer, then:

$$h = \left(\frac{s_2}{s_1} - 1\right)h'$$

h' = differential manometer reading

s_2 = sp. gr. of measuring liquid in differential manometer

s_1 = sp. gr. of liquid flowing in the pipeline

Apparatus: Figure 3.3 shows a venturimeter installed in a pipeline connected to an overhead tank. It consists of a convergent section which reduces in diameter to between one-half and one-fourth of pipe diameter. This is followed by a throat and then a divergent section. The convergent angle is usually 20°.

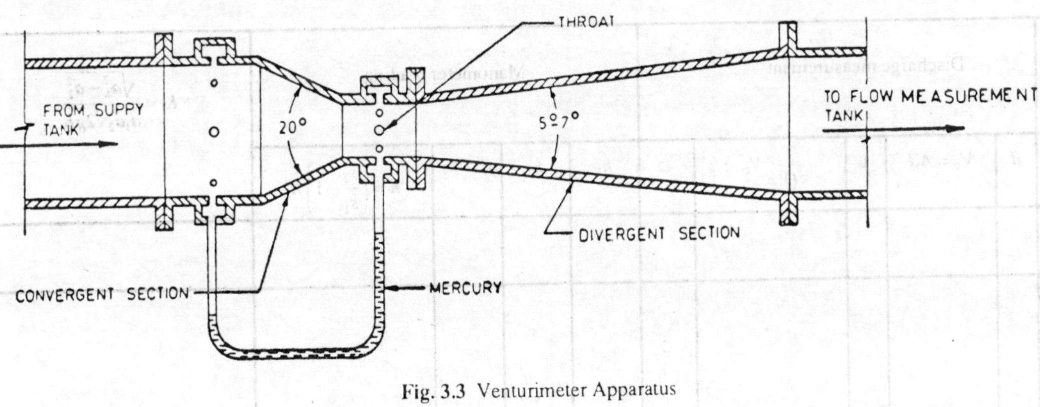

Fig. 3.3 Venturimeter Apparatus

For the divergent part the angle of divergence is usually 5° to 7°. There are two gauge points. One gauge point is at the junction of the pipe and the convergent section and the other gauge point is in the throat portion. Piezometric tappings leading from these gauge points are connected to a differential manometer which indicates the piezometric head. Flow through the venturimeter is controlled with the help of a regulating valve and is collected in a collecting tank having a piezometric tube fixed on a measuring scale.

Experimental Procedure

(i) Open the regulating valve so that water starts flowing through the venturimeter. Wait for some time so that the flow gets stabilized.

(ii) Remove air bubbles, if any, entrapped in piezometric tubes.

(iii) Note differential manometer readings h_1 and h_2.

(iv) Measure the discharge of the apparatus by collecting a certain volume of water in a predetermined time.

(v) Repeat steps (iii) and (iv) for different flow rates and take at least six different sets of observations.

(vi) Take another set of manometer readings for calculation of discharge of the pipeline for constant outflow.

Observations and Calculations

Diameter of the main pipeline, D =

Diameter of the throat section of venturimeter, d =

Area of cross-section of throat section, a_1 =

Area of cross-section of the pipeline, a_2 =

Specific gravity of measuring liquid, s_2 =

Specific gravity of liquid flowing in pipeline, s_1 =

Area of collecting tank, A =

Venturimeter constant, k =

Calculation of Venturimeter Coefficient

S.No.	Discharge measurement				Manometer reading				$K = \frac{\sqrt{a_1^2 - a_2^2}}{a_1 a_2 \sqrt{2gh}}$
	d	$V_0 = Ad$	t	$Q = \frac{V_0}{t}$	h_1	h_2	h'	$h = \left(\frac{s_2}{s_1} - 1\right) h'$	
Units →									
1.									
2.									
3.									
4.									
5.									
6.									

Average value of K =

Calculation of Discharge

S.No.	Manometer reading				$Q = \frac{K a_1 a_2}{\sqrt{a_1^2 - a_2^2}} \times \sqrt{2gh}$
	h_1	h_2	h'	$h = \left(\frac{s_2}{s_1} - 1\right) h'$	
Units					
1.					
2.					
3.					

Average value of Q =

Discussion of Results

Questions for Discussion

(i) How would the results be affected if the venturimeter was not horizontal?

(ii) Would you make any correction to the piezometer readings if the meter were mounted with its axis vertical?

(iii) Why the divergent cone is longer than the convergent cone?

Precautions

(i) There should be no air bubbles entrapped while taking reading of liquid level in piezometric tubes.

(ii) Check that the top level of measuring liquid is same in the two limbs of the differential manometer.

(b) Orifice-Meter

Theory: Orifice-meter, also known as *orifice plate meter*, is a device used to measure the flow rate in any closed pipeline. It is different from the venturimeter in the sense that it provides sudden change in flow conditions instead of smooth transition provided by the venturimeter. As the liquid passes through the orifice-meter, a lot of eddies are formed and there is loss of energy due to which the measured value of discharge, Q, is far less and is given by:

$$Q = C_d \left[\frac{a_1 a_2}{\sqrt{a_1^2 - a_2^2}} \right] \sqrt{2g(h_1 - h_2)} \qquad \text{(iii)}$$

in which C_d is the coefficient of discharge of orifice-meter, a_1 is the cross-sectional area of the pipe and a_2 the cross-sectional area of the orifice as shown in Fig. 3.4. The value of C_d varies from 0.6 to 0.62.

Apparatus: Figure 3.4 shows an orifice-meter installed in a pipeline connected to an overhead tank. It consists of a flat, circular plate having a circular sharp-edged hole called an orifice. The plate is fitted concentrically with the pipe. The diameter of the orifice is about half the diameter of the pipe. There are two gauge points. One gauge point is on the upstream side of the orifice-meter and the other gauge point is just near the orifice-meter on the downstream side. Piezometric tappings leading from the gauge points are connected to a differential manometer which indicates piezometric head. Flow through the orifice-meter is controlled by a regulating valve and is collected in a collecting tank having a piezometric tube fixed on a measuring scale.

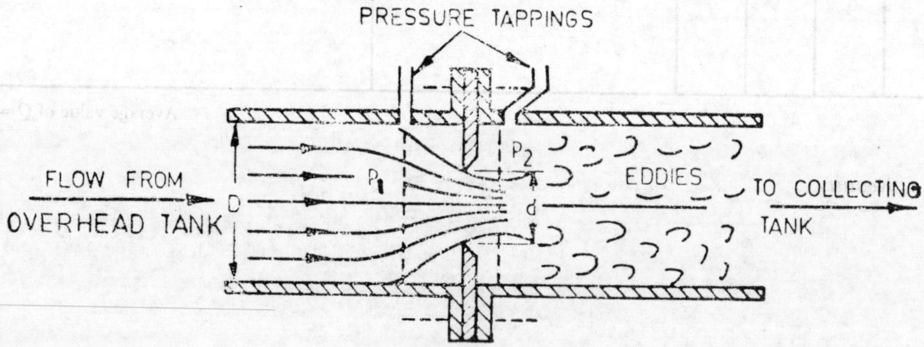

Fig. 3.4 Orifice-meter

Experimental Procedure

(i) Open the regulating valve so that water starts flowing through the orifice-meter. Wait for some time so that the flow gets stabilised.

(ii) Remove air bubbles, if any, entrapped in piezometric tubes.

(iii) Note differential manometer readings h_1 and h_2.

(iv) Measure the discharge by collecting a certain volume of water in a predetermined time.

(v) Repeat steps (iii) and (iv) for different flow rates and take at least six different sets of observations.

(vi) Take another set of manometer readings for calculation of discharge of the pipeline for constant outflow.

Observations and Calculations

Diameter of the main pipe, D =
Area of cross-section of the pipeline, a_1 =
Specific gravity of measuring liquid, s_2 =
Area of collecting tank, A =

Diameter of the orifice, d =
Area of cross-section of the orifice, a_2 =
Specific gravity of liquid flowing in the Pipeline, s_1 =
Coefficient of discharge = C_d

Calculation of Coefficient of Discharge

S.No.	Discharge measurement				Manometer reading				$C_d = Q\dfrac{\sqrt{a_1^2 - a_2^2}}{a_1 a_2 \sqrt{2gh}}$
	d	$V_0 = Ad$	t	$Q = \dfrac{V_0}{t}$	h_1	h_2	h'	$h = \left(\dfrac{s_2}{s_1} - 1\right)h'$	
Units →									
1.									
2.									
3.									
4.									
5.									
6.									

Average value of C_d =

Calculation of Discharge

S.No.	Manometer reading				$Q = \dfrac{C_d a_1 a_2}{\sqrt{a_1^2 - a_2^2}}\sqrt{2gh}$
	h_1	h_2	h'	$h = \left(\dfrac{s_2}{s_1} - 1\right)h'$	
Units →					
1.					
2.					
3.					

Average value of discharge, Q =

Discussion of Results

Question for Discussion

(i) What suggestions do you have for improving the value of the coefficient of discharge?

Precautions: same as for venturimeter apparatus.

Orifice and Mouthpiece

EXPERIMENT 4

(a) To find the value of coefficient of contraction, coefficient of velocity and coefficient of discharge for given:

(i) sharp-edged orifice
(ii) external mouthpiece
(iii) internal mouthpiece

(b) To find the time of emptying a tank when flow takes place through a small sharp-edged orifice.

Theory: Figure 4.1 shows the essential features of flow through an orifice/mouthpiece. The tank is assumed to be sufficiently large for the velocity of flow in it to be negligibly small except close to the orifice. *In the vicinity of the orifice the flow accelerates towards the centre of the hole*, so that as the jet emerges, it suffers a reduction of area due to the curvature of the streamlines, as typified by the streamline *M N* indicated in the figure. The reduction of area due to this local curvature may be taken to be complete at about half the orifice diameter downstream of the plane of the orifice: the reduced section is usually referred to as the vena-contracta.

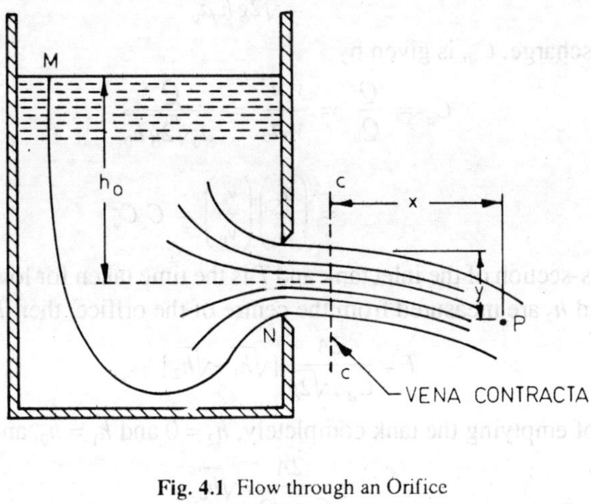

Fig. 4.1 Flow through an Orifice

The coefficient of contraction, C_c, is defined as the ratio of the cross-section of the vena-contracta, a_c, to the cross-section of the orifice, a_0, i.e.,

$$C_c = \frac{a_c}{a_0} \tag{i}$$

Because of the energy loss which takes place as the water passes down the tank and through the orifice, the actual velocity v_c in the plane of vena-contracta will be less than the theoretical velocity, v_0.

The ratio of the actual velocity v_c and the ideal velocity v_o, is often referred to as the coefficient of velocity, C_v of the orifice, i.e.,

$$C_v = \frac{v_c}{v_0} \tag{ii}$$

The theoretical velocity in the plane of the vena-contracta, v_0, can be calculated from the equation.

$$\frac{v_0^2}{2g} = h_0 \quad \text{i.e.,} \quad v_0 = \sqrt{2g\,h_0} \tag{iii}$$

The actual velocity in the plane of the vene-contracta, v_c, is given by the equation

$$v_c = \sqrt{\frac{gx^2}{2y}} \tag{iv}$$

where x and y (measured positive downward) represent the horizontal and vertical coordinates of a point P on the trajectory of the jet (origin being taken at the lowest point of the jet at vena-contracta). Substituting the values of v_c and v_0 in Eq. (ii), we get

$$C_v = \sqrt{\frac{x^2}{4y\,h_0}} \tag{v}$$

Finally, the coefficient of discharge, C_d, is defined as the ratio of the actual discharge to that which would take place if the jet is discharged at the ideal velocity without reduction of area. The actual discharge, Q, given by

$$Q = v_c a_c \tag{vi}$$

and can be measured with the help of measuring tank shown in Fig. 4.2. And if the jet is discharged at the ideal velocity v_0 over the orifice area, a_0, the discharge Q_0 would be

$$Q_0 = v_0 a_0$$

$$= \sqrt{2g\,h_0}\,a_0 \tag{vii}$$

Thus the coefficient of discharge, C_d, is given by:

$$C_d = \frac{Q}{Q_0} = \frac{v_c a_c}{v_0 a_0} = \frac{Q}{a_0\sqrt{2g\,h_0}} \tag{viii}$$

$$= \left(\frac{a_c}{a_0}\right)\left(\frac{v_c}{v_0}\right) = C_c C_v \tag{ix}$$

If A is the area of cross-section of the inlet tank and T is the time taken for lowering the water-level from head h_1 to h_2, where h_1 and h_2 are measured from the centre of the orifice, then T is given by

$$T = \frac{2A}{C_d a\sqrt{2g}}[\sqrt{h_1} - \sqrt{h_2}] \tag{x}$$

For calculating time of emptying the tank completely, $h_2 = 0$ and $h_1 = h_0$, and we get

$$T = \frac{2A}{C_d a\sqrt{2g}}\sqrt{h_0} \tag{xi}$$

Apparatus: Figure 4.2 shows the arrangement of an intake tank which is fed from an overhead tank through a pipe network. Water enters the tank through a regulating valve and leaves through a sharp-edged orifice/mouth piece.

Generally, a revolving plate having various orifices/mouthpieces of different shapes is fixed on one of the walls of the intake tank as shown in the figure. Arrangement is made so that water passes through only one of these openings. The emerging jet is made to fall in the long measuring tank.

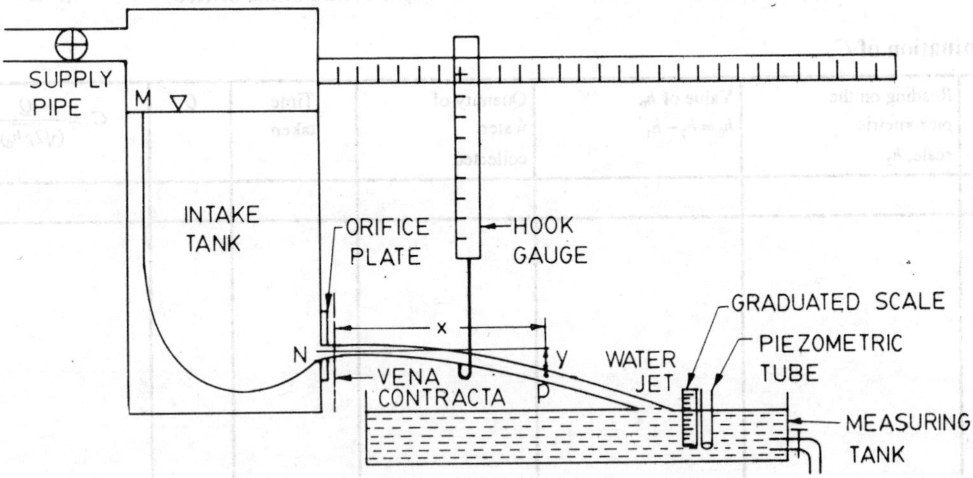

Fig. 4.2 Experimental Set Up for Determination of Coefficients of Orifice

A horizontal scale, on which is mounted a hook gauge, is attached to the intake tank as shown in the figure. Hook gauge can be moved horizontally as well as vertically and its corresponding movements can be read on horizontal and vertical scales respectively.

Experimental Procedure

(i) Rotate the revolving plate so that orifice/mouthpiece of desired shape and size is connected to the opening in the side wall of the intake tank, near its bottom.

(ii) Allow the water to enter the intake tank through the regulating valve and wait till the water-level in the tank becomes steady.

(iii) Measure the head h using a piezometric tube fixed to the intake tank.

(iv) Measure the discharge corresponding to each value of h. Take about eight different flow rates.

(v) Measure x and y coordinates of the lower surface of the jet trajectory at four different points (origin to be taken at lowest point of the jet at vena-contracta).

(vi) Plot graph for Q vs $\sqrt{h}$ for orifice/mouthpiece.

Observations and Calculations

(a) Hydraulic coefficients for sharp-crested orifice/external mouthpiece/internal mouthpiece

(i) Sharp-crested orifice

Shape of the orifice =

Area of cross-section of orifice, a_0 =

Area of cross-section of intake tank, A =

Initial reading on horizontal scale, x_0, and on vertical scale, y_0, when hook gauge just touches the bottom of jet at vena-contracta] x_0 = , y_0 =

Reading on the piezometric scale at the level of the centre of the orifice, h_1 =

Determination of C_d

S.No.	Reading on the piezometric scale, h_2	Value of h_0 $h_0 = h_2 - h_1$	Quantity of water collected	Time taken	Q	$C_d = \frac{Q}{(\sqrt{2g h_0})a_0}$
Units						

Average value of C_d =

(ii) *External mouthpiece*

Shape of the external mouthpiece =

Area of cross-section of external mouthpiece, a_0 =

Area of cross-section of intake tank, A =

Initial reading on horizontal scale, x_0, and on vertical scale, y_0, when hook gauge just touches the bottom of jet at vena-contracta,] x_0 = , y_0 =

Reading on the piezometric scale at the level of the centre of the external mouthpiece, h_1 =

Determination of C_d

S.No.	Reading on the piezometric scale h_2	Value of h_0 $h_0 = h_2 - h_1$	Quantity of water collected	Time taken	Q	$C_d = \frac{Q}{(\sqrt{2g\,h_0})a_0}$
Units						

Average value of C_d =

(*iii*) *Internal mouthpiece*

Shape of the internal mouthpiece
Area of cross-section of internal mouthpiece, a_0 =
Area of cross-section of inlet tank, A =
Initial reading on horizontal scale, x_0, and on vertical scale, y_0, when hook gauge just touches the bottom of jet at vena-contracta] x_0 = , y_0 =
Reading on the piezometric scale at the level of the centre of the internal mouthpiece, h_1 =

Determination of C_d

S.No.	Reading on the piezometric scale h_2	Value of h_0 $h_0 = h_2 = h_1$	Quantity of water collected	Time taken	Q	$C_d = \frac{Q}{a_0\sqrt{2g\,h_0}}$
Units						

Average value of C_d =

Determination of C_v

S.No.	x'	y'	$x = x' - x_0$	$y = y' - y_0$	$C_v = \frac{x}{2\sqrt{y h_0}}$	Remarks

Average value of C_v =

Determination of C_c

$$C_c = \frac{C_d}{C_v} =$$

Also according to Eq. (viii),

$$C_d = \frac{Q}{\sqrt{2g\,h_0}\;a_0}$$

The variation of $\sqrt{h_0}$ with Q should be linear, provided that C_d remains constant over the range of the experiment. Plot Q vs $\sqrt{h_0}$ on an ordinary graph paper and see whether it is true within the limits of experimental error. Find the slope of the straight line and calculate the mean value of C_d from the formula

$$C_d = \frac{1}{a_0\sqrt{2g}}\left(\frac{Q}{\sqrt{h_0}}\right)$$

(*b*) *Time of emptying a tank when flow takes place through a small sharp-edged orifice*

Coefficient of discharge of the orifice, C_d =

Area of cross-section of orifice, a_0 =

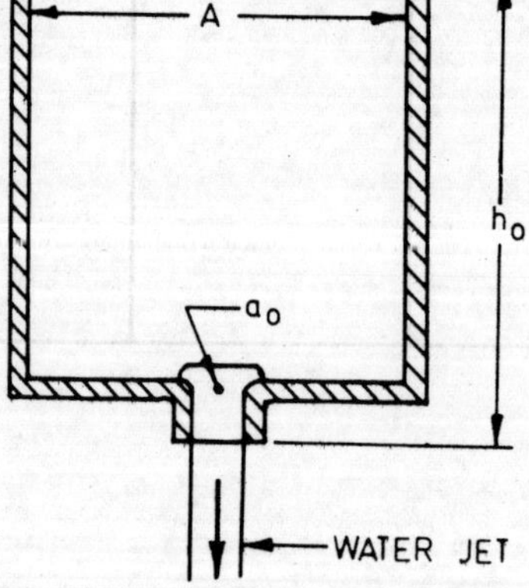

Fig. 4.3 Apparatus Showing Orifice in Bottom of a Tank

Area of cross-section of intake tank, A =

Reading on the piezometric scale at the level of the centre of orifice, h_1 =

S.No.	Initial reading on piezometric scale, h_1	Final reading h_2	$h_0 = h_2 - h_1$	$\sqrt{h_0}$	$T = \dfrac{2A\sqrt{h_0}}{C_d a_0 \sqrt{2g}}$
Units					

Average value of T =

Discussion of Results

Questions for Discussion

(i) Supposing that it is not possible to measure the diameter of the jet at the vena-contracta, but that the diameter is measured some distance away from it, would the results be affected?

(ii) If, when you plot the graph of $\sqrt{h_0}$ against Q, you obtain a line which does not appear to pass through the origin, what possible reasons are there for this? How will the value of C_d obtained from the graph be affected.

Precautions

(i) When taking reading with the piezometric tube, check that air is not trapped in this tube.
(ii) Take care that water-level in the inlet tank is fairly constant when taking a reading.
(iii) The apparatus should be drained after use.

Flow Through Pipes

EXPERIMENT 5

To find out the critical velocity of flow by Reynolds experiment

Theory: Osborne Reynolds, in 1883, conducted a number of experiments to determine the laws of resistance in pipes. By introducing a filament of dye into the flow of water along a glass pipe, he showed the existence of two different types of motion. At low velocities the filament appeared as a straight line which passed down the whole length of the tube, indicating laminar flow. At higher velocities, the filament, after passing a little way along the tube, suddenly mixed with the surrounding water, indicating that the motion had now become turbulent.

Experiments with pipes of different diameters and with water at different temperatures led Reynolds to conclude that the parameter which determines whether the flow shall be laminar or turbulent in any particular case is:

$$\mathbf{Re} = \frac{\rho V D}{\mu} \qquad \text{(i)}$$

in which **Re** denotes the Reynolds Number of motion

ρ denotes the density of fluid

V denotes the velocity of flow

D denotes the diameter of the pipe

μ denotes the coefficient of viscosity of the fluid

The motion is laminar or turbulent according as the value of **Re** is less than or greater than a certain value. If experiments are made with increasing rate of flow, the value of **Re** depends on the degree of care which is taken to eliminate disturbances in the supply or along the pipe. On the other hand, if experiments are made with decreasing flow, transition from turbulent to laminar flow takes place at a value of **Re** which is very much less dependent on initial disturbances. This value of **Re** is about 2000 for flow through circular tubes, and below this the flow is inherintly laminar in nature. The velocity at which the flow in the pipe changes from one type of motion to the other is known as critical velocity and is given by:

$$V = \frac{\mathbf{Re}\,\mu}{\rho D} \qquad \text{(ii)}$$

Also, the value of critical velocity corresponding to **Re** = 2000, also known as lower critical Reynolds Number, is called Lower Critical Velocity. The upper critical Reynolds Number (the maximum value of **Re** at which laminar flow is physically possible) depends largely upon the nature of disturbance present in the flow and may be as high as 40,000.

Apparatus: In Fig. 5.1 is shown the Reynolds' apparatus. A glass tube with a bell mouthed inlet leads from a large tank and terminates in a regulating valve by which the rate of flow can be controlled. The water is made to pass through the holes of the slotted plates fixed in the tank, so as to reduce the velocity with which the water approaches the bell-mouthed entrance of the glass tube. The bell-mouthed entrance provides a passage to ensure streamlined flow. A dye container is attached to the side wall of the tank injecting a stream of dye into the glass tube.

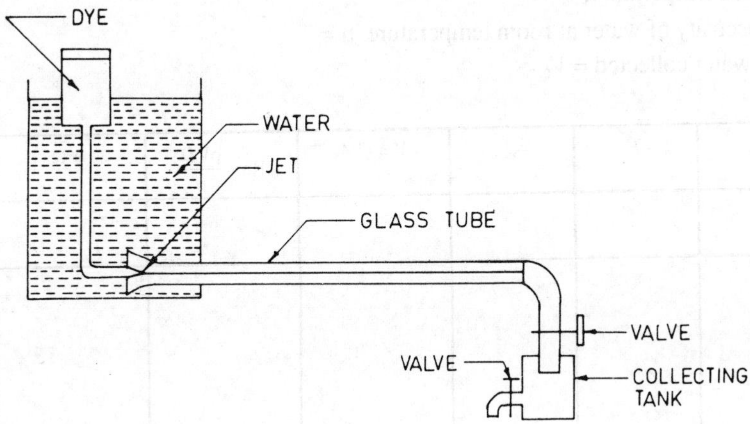

Fig. 5.1 Reynolds' Apparatus

Experimental Procedure

(i) Open the main supply valve and fill the tank of the apparatus with water; with outlet of glass tube partly open so that no air is entrapped in the glass tube.

(ii) Close the outlet valve of the glass tube and the inlet valve of the tank when the tank is full.

(iii) Leave the apparatus for some time so that water in the tank is at rest. No disturbance.

(iv) Partially open the outlet valve of the glass tube and inlet of the tank so that velocity of flow is very small and the water-level in the tank is fairly constant.

(v) Open inlet of the dye-injector so that the dye stream moves as a straight line through the tube showing that the flow is laminar.

(vi) Increase the velocity of flow and again measure the discharge.

(vii) Take six to eight readings till the dye stream in the glass tube breaks up and gets diffused in water. Also encircle the reading for which the dye filament wavers for the first time near the outlet end of glass tube.

(viii) Note down the room temperature at least three times during the experiment.

(ix) Repeat the experiment with decreasing rate of flow and encircle the reading for which dye filament wavers for the last time near outlet end of glass tube; as the flow changes from turbulent to laminar.

Observations and Calculations

Inner diameter of glass tube, $D =$

Cross-sectional area of glass tube, $A = \frac{\pi}{4} D^2$

Mass density of water, $\rho =$

Average room temperature, $\theta =$

Dynamic viscosity of water at room temperature, $\mu =$

Volume of water collected $= V_0$

S.No.	V_0	t	Q	$V = Q/A$	$\text{Re} = \frac{\rho V D}{\mu}$	θ°C	Remarks
Units →							
1.							
2.							
3.							
4.							
5.							
6.							

∴ Critical velocity of flow =

Discussion of Results

Precautions

(i) Do not forget to remove any entrapped air in the apparatus before starting measurement.
(ii) There should be no mechanical vibration near the apparatus.
(iii) Do not forget to record the temperature of water at frequent intervals.
(iv) Increase in velocity of flow should be in stages.

EXPERIMENT 6

(a) To determine Darcy's friction factor 'f' of pipes of different diameters.
(b) To verify the losses of head due to
  (i) friction,
  (ii) sudden enlargement,
  (iii) sudden contraction, and
  (iv) right-angled bend.

Theory: While the nature of flow depends upon the Reynolds Number, the frictional resistance offered to the flow of fluids depends essentially on the roughness of the surface of the conduit carrying the flow. In laminar flow this frictional resistance is mostly due to viscous resistance of fluid to flow. In turbulent flow it is due to resistance offered by viscosity of fluid and surface roughness of the conduit.
Experiments have shown that the frictional resistance varies
  (i) with the degree of roughness of surface with which fluid comes in contact,
  (ii) with the extent of area of surface coming in contact with fluid,
  (iii) directly as the velocity in laminar flow and as square of velocity in turbulent flow,
  (iv) directly as the density of the fluid, and
  (v) inversely as the viscosity of fluid.

This frictional resistance causes loss of head, h_f, which is given by Darcy and Weisbach equation:

$$h_f = f\left(\frac{L}{D}\right)\frac{v^2}{2g} \qquad \text{(i)}$$

where f is called Darcy's friction factor and is given by $f = \frac{64}{\mathbf{Re}}$ for laminar flow and depends upon relative roughness of pipe in case of turbulent flow. But the head losses due to change of section, bends, elbows, valves and fittings of all types are classified as minor losses in pipelines.

Minor losses usually result from rather abrupt changes (in magnitude and direction) of velocity. In general, increase of velocity (acceleration) is associated with small head loss but decrease of velocity (deceleration) causes large head loss because of the production of large-scale turbulence.

Early experiments with water (at high Reynolds Number) indicated that minor losses vary approximately with the square of velocity and lead to the proposal of the basic equation

$$h_L = K\frac{v^2}{2g} \qquad \text{(ii)}$$

in which K, the loss coefficient, is practically constant.
Eq. (i) can also be written as

$$h_f = K\frac{v^2}{2g}$$

$$\text{where } K = \frac{fL}{D}$$

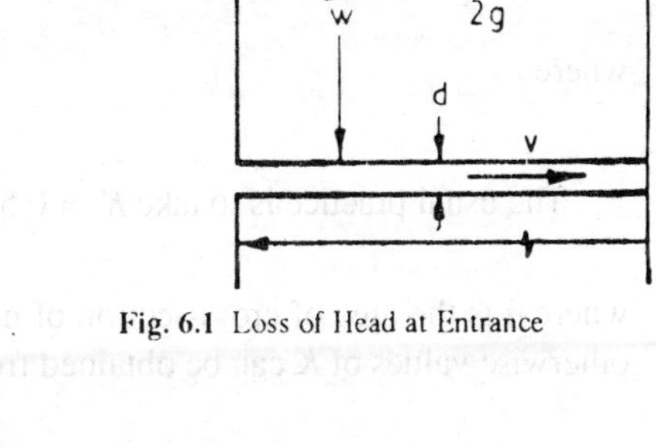

Fig. 6.1 Loss of Head at Entrance

(i) Loss of head at entry: It is given by the formula

$$h_L = 0.5\frac{v^2}{2g} \qquad \text{(iii)}$$

where v is the velocity of flow in the pipeline coming out of reservoir as shown in Fig. 6.1.

(*ii*) *Loss of head due to sudden enlargement:* It is given by the formula

$$h_L = \frac{(v_1 - v_2)^2}{2g} \qquad \text{(iv)}$$

where v_1 and v_2 are velocities of flow as shown in Fig. 6.2.

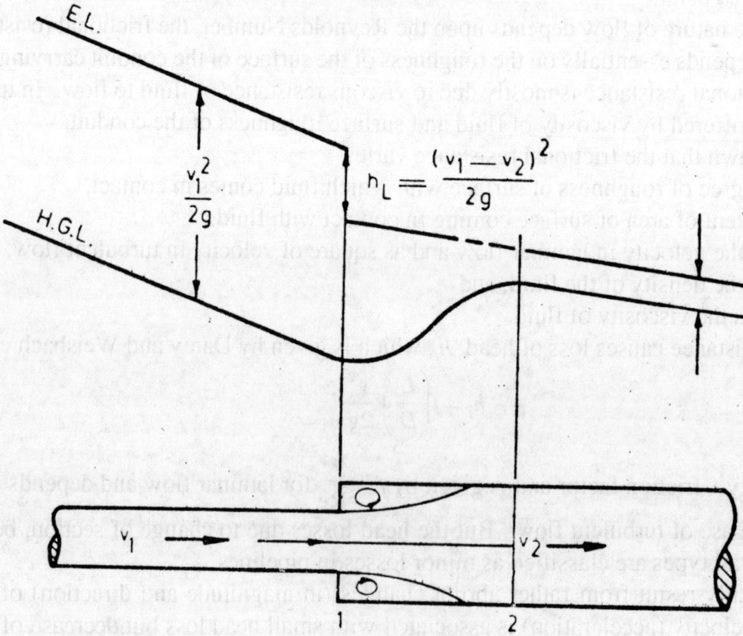

Fig. 6.2 Sudden Enlargement

(*iii*) *Loss of head due to sudden contraction:* It is given by the formula

$$h_L = \left(\frac{1}{C_c} - 1\right)^2 \frac{v_2^2}{2g} \qquad \text{(v)}$$

$$= K\frac{v_2^2}{2g}$$

where

$$K = \left(\frac{1}{C_c} - 1\right)^2$$

The usual practice is to take $K = 0.5$ for all values of $\left(\frac{a_1}{a}\right)$

where a is the area of cross-section of main pipe and a_1 is the area of cross-section of contracted section, otherwise values of K can be obtained from Table 6.1.

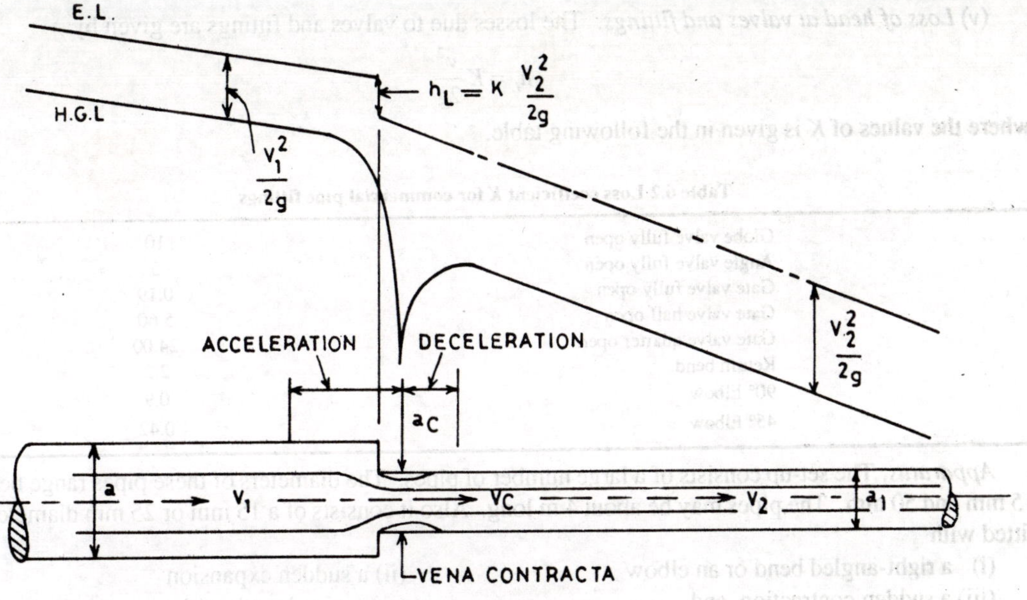

Fig. 6.3 Sudden Contraction

Table 6.1

a_1/a	0	.1	.2	.3	.4	.5	.6	.7	.8	.9	1
C_c	.617	.624	.632	.643	.659	.681	.712	.755	.813	.892	1
$(1/C_c - 1)^2$	.385	.365	.340	.308	.266	.219	.164	.106	.055	.015	0

(iv) *Loss of head at exit:* It is given by the formula

$$h_L = \frac{v^2}{2g}$$

where v is the velocity of flow in the pipe near the exit.

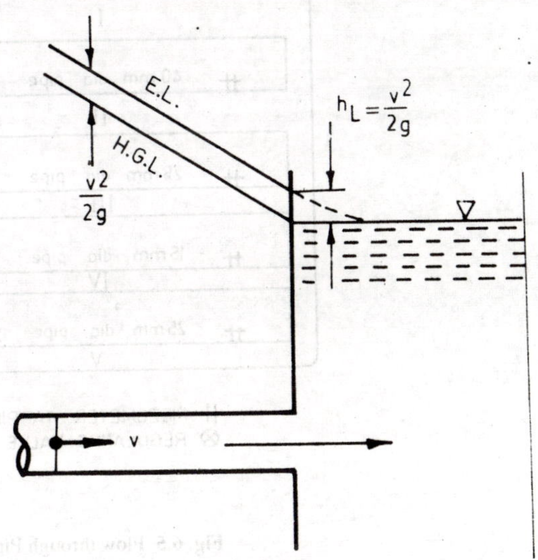

Fig. 6.4 Pipe Exit

(v) ***Loss of head at valves and fittings:*** The losses due to valves and fittings are given by

$$h_L = K \frac{v^2}{2g}$$

where the values of K is given in the following table.

Table 6.2 Loss coefficient *K* for commercial pipe fittings

Globe valve fully open	10
Angle valve fully open	5
Gate valve fully open	0.19
Gate valve half open	5.60
Gate valve quarter open	24.00
Return bend	2.2
90° Elbow	0.9
45° Elbow	0.42

Apparatus: The set-up consists of a large number of pipes. The diameters of these pipes range between 15 mm and 50 mm. The pipes may be about 4 m long. Also it consists of a 15 mm or 25 mm diameter pipe fitted with

(i) a right-angled bend or an elbow
(ii) a sudden expansion
(iii) a sudden contraction, and
(iv) a pipe length of 4 m

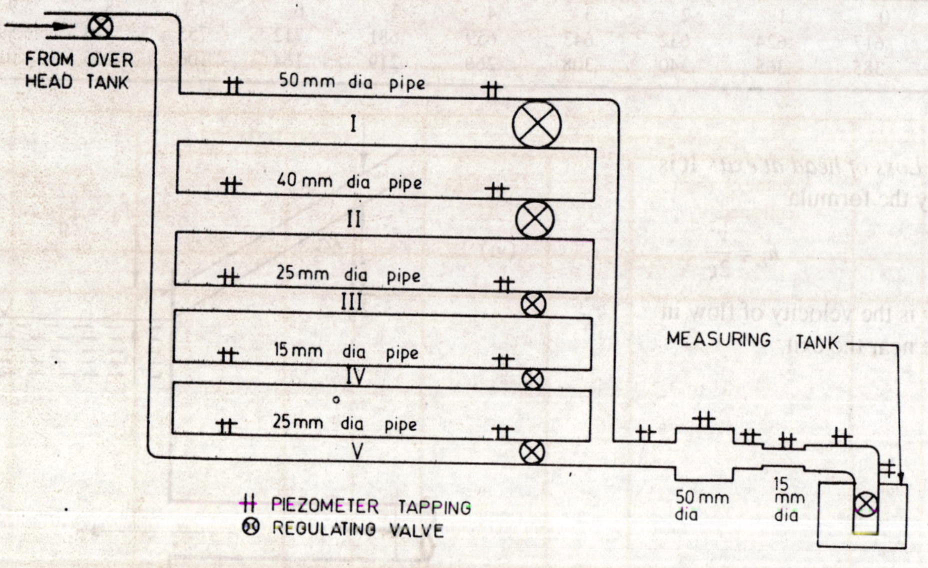

Fig. 6.5 Flow through Pipe/Pipe Fittings Apparatus

Flow to the apparatus is regulated by means of a regulating valve provided near the outlet end of each pipe as shown in Fig. 6.5. One common inlet valve is provided in the main supply line for the whole apparatus. For measuring the head loss, two pressure tappings are provided on either side of the pipe fittings and pressure tapping along the pipe may be 3 m apart. The tapping may be connected to a manometer turn by turn. Sufficient length of the pipeline is provided between various fittings. Discharge of the apparatus is measured in a measuring tank.

(a) Determination of Co-efficient of Friction of Pipes of Different Diameters

Experimental Procedure

(i) Connect the manometer rubber tubings to gauge points of one of the pipelines.
(ii) Open the inlet valve, keeping the outlet valve closed.
(iii) Check if there is any air bubble in the manometer tube in which case some pressure difference will be existing between gauge points to which manometer is connected. Remove air bubble, if any.
(iv) Open partially the outlet valve, keeping the common inlet valve fully open.
(v) Allow the flow to get stabilised and then take manometer reading.
(vi) Measure the discharge.
(vii) Repeat steps (iv) to (vi) for six different discharges.
(viii) Repeat steps (iv) to (vii) for other pipes.

Observations and Calculations

Particulars	Pipe I	Pipe II	Pipe III	Pipe IV
Pipe diameter, D	50 mm	40mm	25mm	15mm
Distance between gauge points, L				

Pipe No.	S. No.	Discharge measurement			Manometer reading				
		Volume of water collected, V_o	Time t	Discharge Q	h_1	h_2	$h_f = h_1 - h_2$	$f = \dfrac{2gDh_fA^2}{LQ^2}$	Average f
Units									
I (50 mm dia.)									

(Contd.)

II (40 mm dia.)								
III (25 mm dia.)								
IV (15 mm dia.)								

Discussion of Results

(b) Calculation of Various Losses

Experimental Procedure

(i) Open the inlet valve, keeping the outlet valve closed.
(ii) Connect the manometer rubber tubings to one of the pipes/pipe fittings and check that there is no air bubble entrapped.
(iii) Open partially the outlet valve, keeping the common inlet valve fully open.
(iv) Allow the flow to get stabilised and then take manometer reading.
(v) Measure the discharge.
(vi) Take at least four readings.
(vii) Repeat steps (i) to (vi) for different pipe fittings.

Observations and Calculations

Diameter of main pipe, D =
Diameter of enlarged pipe, D_1 =
Area of cross-section of main pipe, a =
Area of cross-section of enlarged pipe, a_1 =
Length of main pipe between pressure tapping, L =

Pipe/Pipe fitting	Run No.	Discharge measurement			Velocity of flow in			h_f/h_L	
		Volume of water collected, V_o	Time t	Discharge Q	Main pipe $v = Q/a$	Enlarged pipe $v_1 = Q/a_1$	K	Calculated	Measured
	⟶ Units								
Pipe									
Bend									
Sudden enlargement									
Sudden contraction									

Discussion of Results

Precautions

(i) Take care that there is no air bubble entrapped in the apparatus when noting manometer reading.
(ii) There should be no leakage from any of the pipe fittings.

EXPERIMENT 7

To study and use Moody's Diagram, Nomograph of Manning's Equation and Nomograph used for water-supply pipes and drainage sewers.

Theory:

Flow Under Pressure in Circular Pipes: Scientific speculations concerning the physical relationships controlling pipe flow date back to the middle of the 19th century. Fundamental concepts were laid down by Chezy (1718-1798) whose formula

$$v = C\sqrt{mi} \tag{i}$$

is also used for open-channel flow. But Darcy and Weisbach (1806-1871) gave more satisfactory approach. Their formula

$$h_f = f\frac{L}{D}\left(\frac{v^2}{2g}\right)$$

states that the head loss in a given pipe is proportional to the kinetic energy, $\frac{v^2}{2g}$ of flow.

In case of laminar flows the friction factor is computed from the formula

$$f = \frac{64}{\mathbf{Re}} \tag{ii}$$

showing a linear relationship between Reynolds Number and the friction factor. But in case of turbulent flow the influence of the roughness of the pipe wall becomes dominant. The roughness size e is expressed in terms of pipe diameter as:

$$\in = \frac{e}{D} \tag{iii}$$

where $\in$ is called the relative roughness.

The first extensive determination of the effect of pipe roughness upon the flow was done by Nikuradse. The results of Nikuradse's experiment indicated that when the roughness of the pipe fully influences the flow at high Reynolds Number, the friction factor f is independent of the **Re**. Von Karman derived a formula for turbulent flow within rough pipe boundaries and the resulting equation for the value of friction factor is

$$\frac{1}{\sqrt{f}} = 1.14 + 2\log\frac{D}{e} \tag{iv}$$

Other investigators developed additional formulas for expressing Nikuradse's experimental results in mathematical forms. Lewis F. Moody (1880-1953) has plotted these equations on a graph similar to one reproduced in Fig. 7.1 called Moody's Diagram. Both experimental and theoretical investigations lead to uncertain results in the transitional range when the flow changes from laminar to turbulent conditions. But from a practical point of view this is rather immaterial because almost all designs concern flow well within the turbulent range. With Darcy Weisbach equation and with Moody's diagram available, frictional loss can be computed with relative ease and reasonable accuracy.

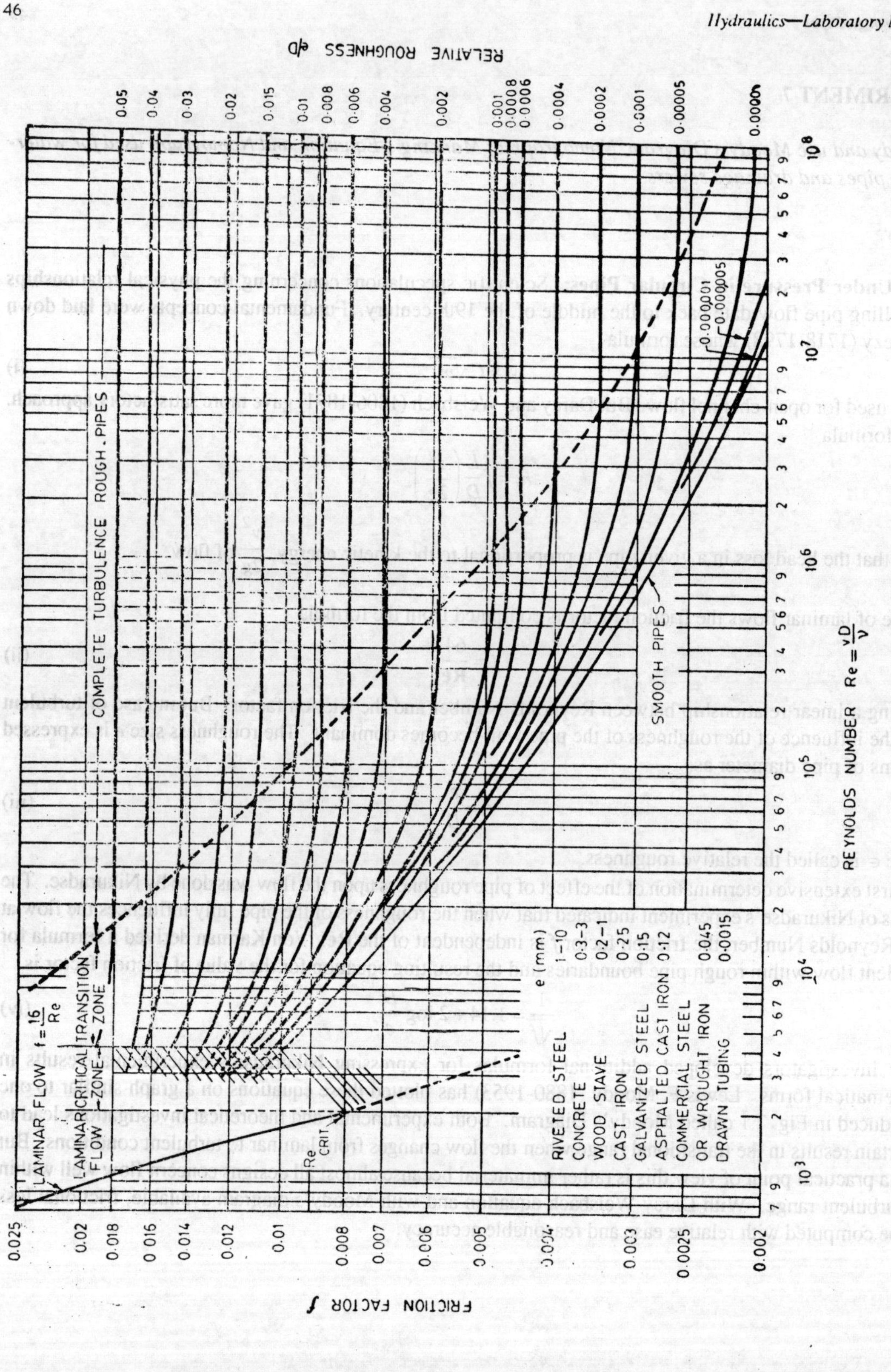

Fig. 7.1 Moody's Diagram

But recent surveys indicate that as much as 80 per cent of engineers in water-supply practise prefer to use old Hazen William's Nomograph (Fig. 7.2) instead of Darcy-Weisbach equation because of the following two reasons:

(i) The roughness of the pipe is still unknown, particularly if we consider the long time in which a pipeline is in use; due to which design cannot be very accurate.

(ii) Cumbersome trial-and-error procedure to be followed in case Q, the discharge, is unknown.

Hazen William's formula, most commonly used in the design and evaluation of a water distribution system, is:

$$v = 0.849 C m^{0.63} i^{0.54} \quad \text{(v)}$$

where C = coefficient that depends upon the material and age of the conduit given in Table 7.1
m = hydraulic radius,
i = slope of hydraulic gradient, m/m
v = velocity of flow in m/s

Since

$$v = \frac{Q}{A} = \frac{4Q}{\pi D^2} \quad \text{(vi)}$$

and

$$m = \frac{A}{P} = \frac{(\pi/4)D^2}{\pi D} = \frac{D}{4} \quad \text{(vii)}$$

where A = cross-sectional area,
P = wetted perimeter,
D = diameter, in metre, of the circular pipe flowing full

Eq. (v) can be modified as

$$Q = 0.278 C D^{2.63} i^{0.54} \quad \text{(viii)}$$

The nomograph shown in Fig. 7.2 solves Eq. (viii) for a coefficient C equal to 100.

Table 7.1 Values of coefficient C for Hazen William's formula

Pipe material	C
Asbestos cement	140
Cast iron cement lined	130 to 150
New unlined	130
5 year old, unlined	120
20 year old, unlined	100
Concrete	130
Copper	130 to 140
Plastic	140 to 150
New welded steel	120
New riveted steel	110

Head losses in pipes with coefficient values other than 100 can be determined by using the correction factor, K, given in Table 7.2 where corrected $h_f = K$ (h_f at C = 100).

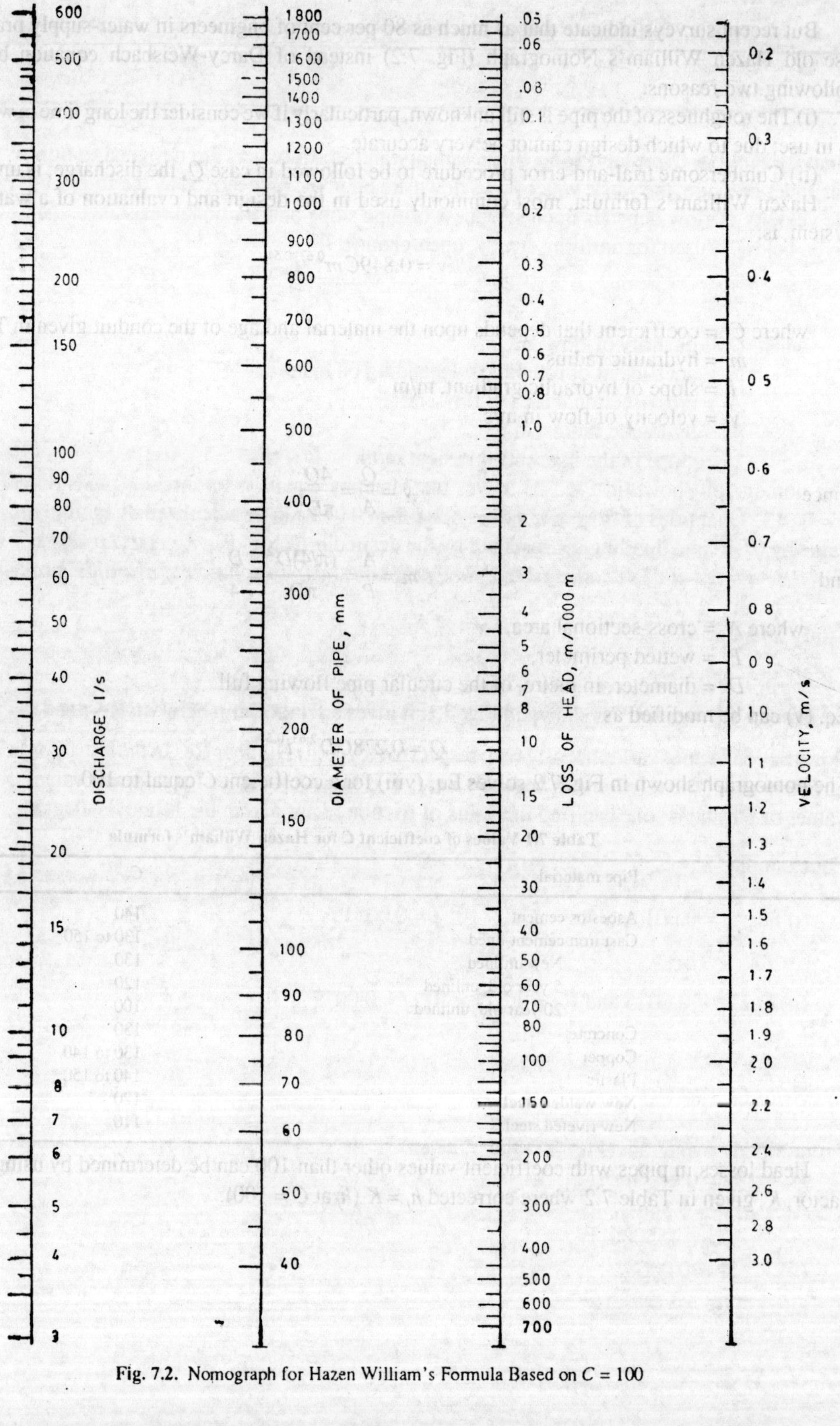

Fig. 7.2. Nomograph for Hazen William's Formula Based on $C = 100$

Table 7.2 Correction factor to determine head losses from Fig. 7.2 at values of *C* other than *C* = 100

C	80	100	110	120	130	140
K	1.51	1.00	0.84	0.71	0.62	0.54

Gravity Flow in Circular Pipes: Since sanitary and storm sewers are designed to flow as open channels, not under pressure, the waste water flows downstream in the pipe moved by the force of gravity. Also since the velocity of flow depends upon steepness of pipe slope and frictional resistance, the Manning formula, given below, is used for uniform, steady, open-channel flow

$$Q = \frac{1.00}{n} A m^{2/3} i^{1/2}$$

where Q = quantity of flow, m^3/s
n = coefficient of roughness depending on material
A = cross-sectional area of flow, m^2
m = hydraulic radius, m
i = slope of the hydraulic gradient m/m

The nomograph shown in Fig. 7.3 solves the Manning equation for circular pipes flowing full based on $n = 0.013$. Quantities of flow at n-values other than 0.013 can be calculated by multiplying the nomograph value by 0.013 and dividing the resultant by the desired n-factor. But *for circular pipes flowing partly full, Fig. 7.3 along with 7.4, called partial flow diagrams, help in solution of hydraulic problems.*

Study and Use of

(a) Moody's Diagram: As shown in Fig. 7.1, it gives a relationship between **Re**, f and $\in = \frac{e}{D}$. For laminar flow friction factor f varies linearly with respect to Reynolds Number **Re**. In the transition zone and complete turbulent zone, the relationship between **Re**, f, $\in$ is given by curves for different values of f. Knowing the values of **Re** and $\in$, one can find the value of friction factor f from the Moody's diagram.

For example,

(i) For $\frac{e}{D} = 0.001$ and **Re** $= 3.72 \times 10^5, f = 0.00515$

(ii) For $\frac{e}{D} = 0.0025$ and **Re** $= 1.773 \times 10^4, f = 0.0079$

Problems for Practice

From Moody's diagram (Fig. 7.1) find f, when

(i) $e = 0.003$ m
$D = 0.3$ m
Re = 455000 f = ________

(ii) $\in = 0.0008$
Re $= 5.08 \times 10^5$ f = ________

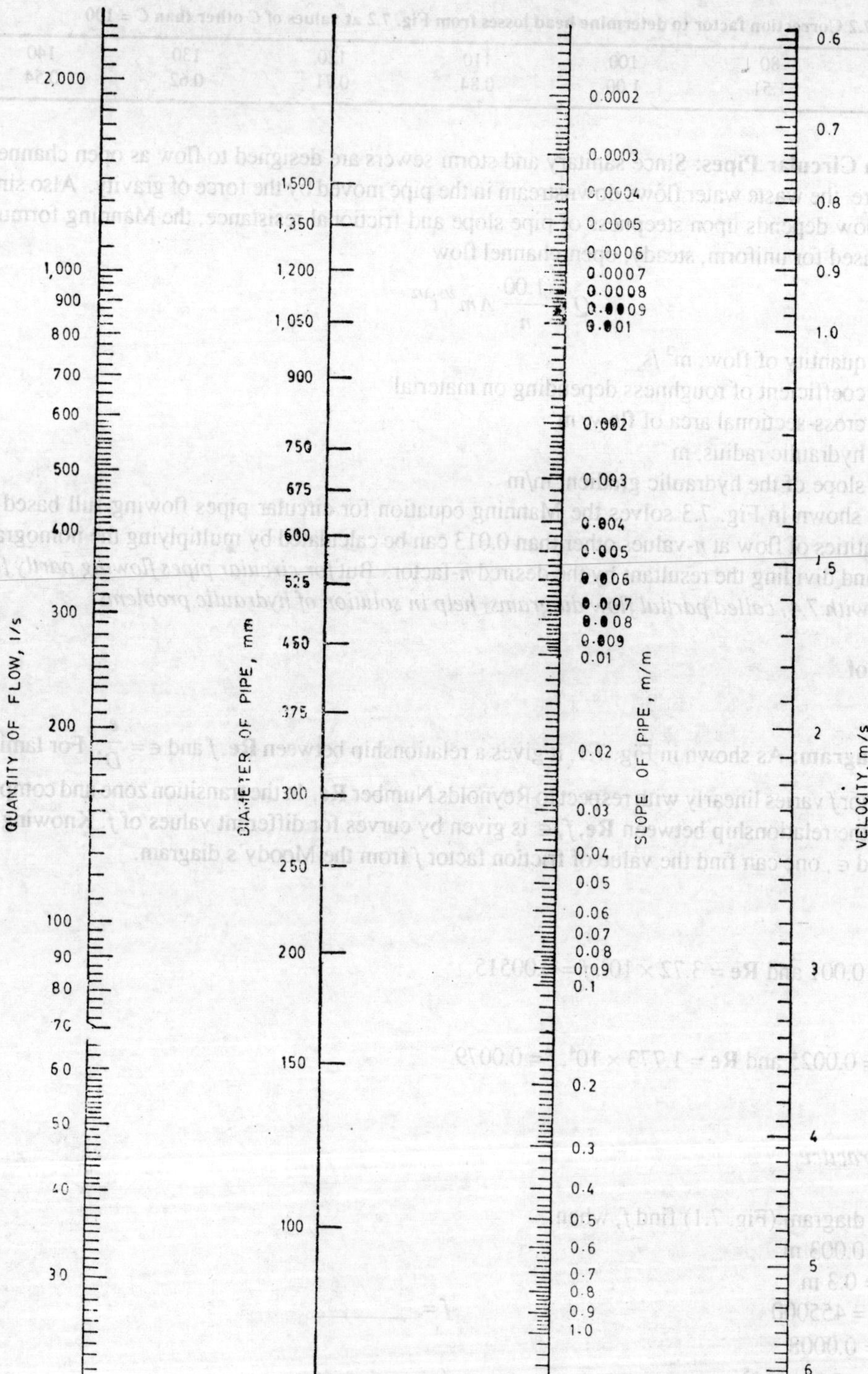

Fig. 7.3 Nomograph for Manning Formula for Circular Pipes Flowing Full Based on $n = 0.013$

(b) Nomograph for Hazen William's Formula: As shown in Fig. 7.2, this nomograph gives solution of Hazen William's formula

$$Q = 0.278CD^{2.63}i^{0.54}$$

for $C = 100$. Given any two of the parameters (discharge, diameter of the pipe, loss of head, or velocity), the remaining two can be determined from the intersections along a straight line drawn across the nomograph.

For example:
'the flow of 30 l/s in a 200 mm diameter pipe has a velocity of 0.96 m/s with a head loss of 8.1 m/1000 m for $C = 100$'.
Head losses in pipes with coefficient values other than 100 can be determined by using the correction factor given in Table 7.2.
For example, if the head loss at $C = 100$ is 8.1 m/1000 m, the head loss at $C = 130$ would be equal to $0.62 \times 8.1 = 5.0$ m/1000 m.

Problems for Practice

From nomograph for Hazen William's formula,

(i) if D = 40 mm, C = 100
v = 0.6 m/s
Find Q = ________ and h_f = ________.

(ii) if Q = 30 l/s v = 0.6 m/s, C = 100
Find D = ________ and h_f = ________.

(c) Nomograph for Manning's Formula for Circular Pipes Flowing Full for $n = 0.013$: As shown in Fig. 7.3, this nomograph gives solution of Manning's equation for circular pipes flowing full based on a coefficient of roughness of 0.013. Given any two of the parameters (quantity of flow, diameter of pipe, slope of pipe, or velocity) the remaining two can be determined from the intersection along a straight line drawn across the nomograph.

For example a 200 mm diameter pipe set on a slope of 0.02 mm conveys a quantity of flow of 46 l/s flowing full with an average velocity of 1.5 m/s.

Quantities of flow at n-values other than $n = 0.013$, can be calculated by multiplying the nomograph value by 0.013 and dividing the resultant by the desired n factors.

Problems for Practice

From nomograph for Manning's formula,

(i) if D = 250 mm, for $n = 0.013$
i = 0.01
then Q = ________ and v = ________
but for n = 0.015
Q = ________ and v = ________

(ii) if D = 450 mm for $n = 0.013$
i = 0.0025
then Q = ________ and v = ________

(d) Circular Pipes Flowing Partly Full: Hydraulics problems of circular pipes flowing partly full are solved by using Fig. 7.3 along with Fig. 7.4. Figure 7.4, called partial flow diagram, gives ratio of hydraulic elements such as quantity of flow, area of flow and velocity of flow with respect to ratio of depth of flow to diameter of pipe. The symbols *q*, *v*, *a* relate to the partial-flow conditions while *Q*, *V*, *A* to full-flow conditions.

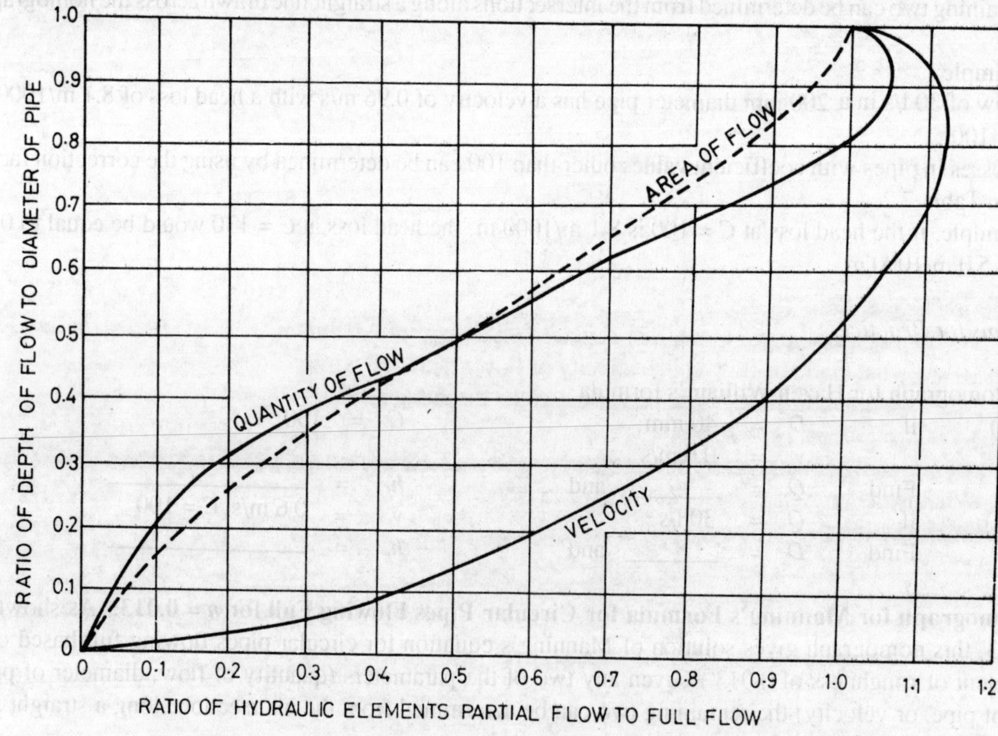

Fig. 7.4 Relative Quantity, Velocity and Cross-sectional Area of Flow in a Circular Pipe for any Depth of Flow

Consider flow in a pipe at a depth of 30 per cent of pipe diameter represented by the horozontal line labelled 0.3. Where this line intersects the curved lines for quantity of flow, area of flow and velocity, project it downward to the horizontal scale and read respectively, the values 0.2, 0.25 and 0.78; these values mean that partial flow at this depth conveys a quantity of flow equal to 20 per cent of flowing full quantity, the cross-sectional area of flow is 25 per cent of total area of the pipe and the average velocity of flow is 78 per cent of the flowing full velocity.

Example: *The measured depth of flow in a 1200 mm concrete storm sewer on a grade of 0.0015 m/m is 740 mm. What is the calculated quantity and velocity of flow?*

Solution: From Fig. 7.3, for i = 0.0015 m/m and D = 1200 mm, we find that Q = 1500 l/s and V = 1.34 m/s.

Depth of flow to diameter of pipe ratio is

$$\frac{d}{D} = \frac{740}{1200} = 0.62$$

from Fig. 7.4, for $\frac{d}{D} = 0.62$ we find that

$$\frac{q}{Q} = 0.72 \text{ and } \qquad \frac{v}{V} = 1.08$$

for depth of flow, $d = 740$ mm, we find

$$Q = 0.72 \times 1500 = 1100 \text{ l/s}$$

and $\quad v = 1.08 \times 1.34 = 1.45$ m/s

Problems for Practice

1. A 750 mm sewer pipe ($n = 0.013$) is placed on a slope of 4.0 mm/m.
   (i) At what depth of flow is the velocity of flow equal to 0.6 m/s.
   (ii) If the depth of flow is 450 mm, what is the discharge?
   [**Ans.** (i) 75 mm (ii) 470 l/s]
2. A 1500 mm diameter storm sewer conveys a flow of 1.5 m^3/s at a depth of flow equal to 1000 mm. What is the unused capacity of the pipe in m^3/s?
   [**Ans.** 0.4 m^3/s]

Flow Measurements

EXPERIMENT 8

(*a*) *To demonstrate the use of Pitot tube.*
(*b*) *To determine velocity distribution in a given pipeline.*

(a) Demonstration of the Use of Pitot Tube

Theory: Velocity measurements in pipes, as well as in open channels, may be made by measuring the pressure corresponding to the kinetic energy of flow. The simplest method is to measure the pressure in an open-ended tube bent in such a way that the end is aligned opposite to the velocity vector measured. Such tubes are called Pitot tubes. Invented by a French scientist, Henry Pitot (1695-1771), in 1732, this tube was first used to measure the flow of river Seine in Paris. The kinetic energy at the centre of this tube converts into pressure energy as the flow of fluid is halted at that point as shown in Fig. 8.1.

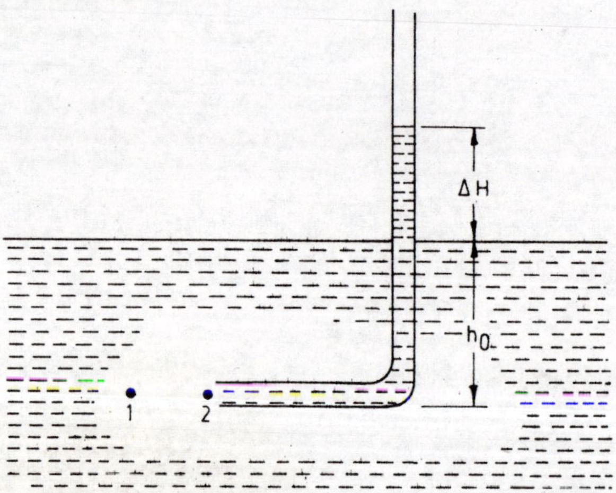

Fig. 8.1 Simple Pitot Tube

The tube opening is directed upstream so that the fluid flows into the opening until the pressure builds up in the tube sufficiently to withstand the impact of velocity against it. Directly in front of the opening, the fluid is at rest.
With reference to Fig. 8.1, Bernoulli's equation applied between points 1 and 2 gives

$$\frac{v^2}{2g}+\frac{p_1}{w}=\frac{p_2}{w}$$

since both points are at the same elevation and velocity at point 2 is zero.

i.e.,
$$\frac{v^2}{2g}+\frac{p_1}{w}=h_o+\Delta h \qquad \text{(i)}$$

As $\frac{p_1}{w}=h_0$, Eq. (i) is reduced to

$$\frac{v^2}{2g}=\Delta h \text{ or } v=\sqrt{2g\,\Delta h} \qquad \text{(ii)}$$

For an open stream of liquid, only this single tube is necessary since the difference between stagnation pressure and static pressure is measured directly. But for an enclosed stream of liquid, or for a gas, the Pitot tube indicates simply the stagnation pressure and so the static pressure must be measured separately.

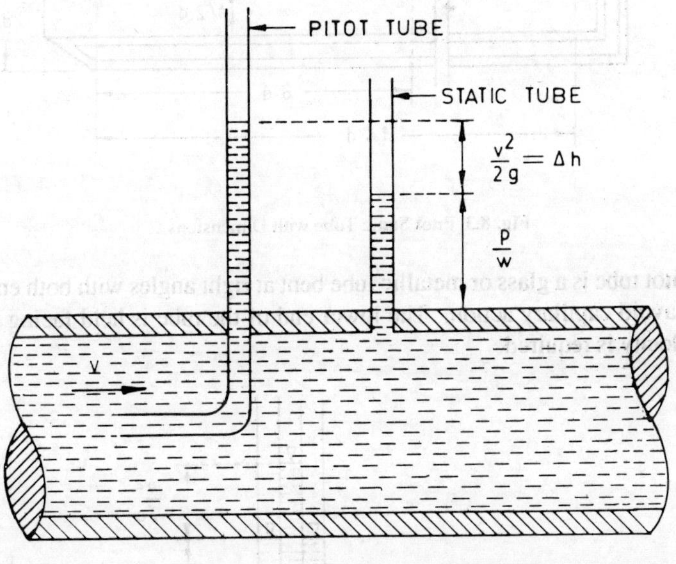

Fig. 8.2 Pitot Tube along with Static Tube

In pipes the pressure difference Δh between a Pitot tube and a nearby static tube provides a means to compute the velocity at the point, measured as shown in Fig. 8.2,

$$v=\sqrt{2g\,\Delta h} \qquad \text{(iii)}$$

where
$$\Delta h=\frac{\Delta p}{w}$$

The static tube and Pitot tube may be combined into one instrument called a pitot-static tube, as shown in Fig. 8.3. Prandtl tube is a particular form of pitot-static tube with blunt nose. Standard proportions of Pitot static tube have been determined which gives very accurate results. If other proportions are used, a correction factor C determined by calibration has to be introduced and $v = C\sqrt{2g\,\Delta h}$ is modified expression for velocity of flow.

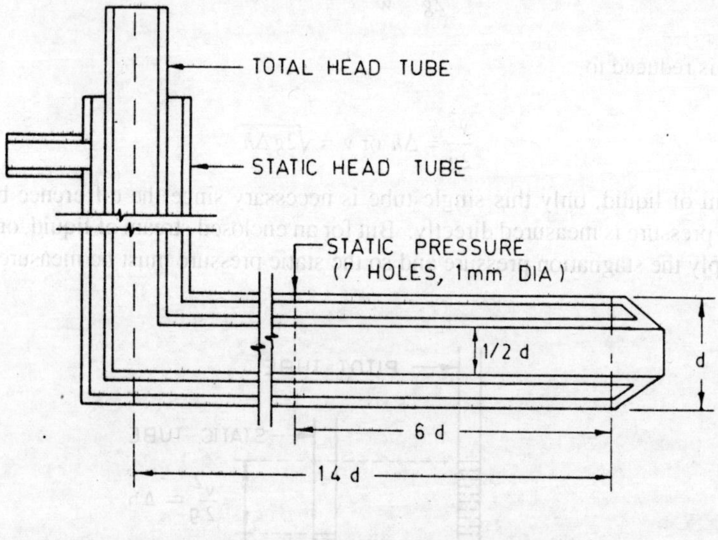

Fig. 8.3 Pitot Static Tube with Dimensions

Description: A Pitot tube is a glass or metallic tube bent at right angles with both ends open and having adequate diameter to avoid capillary action. The lower end of the tube is held facing the current and at a depth at which the velocity is required.

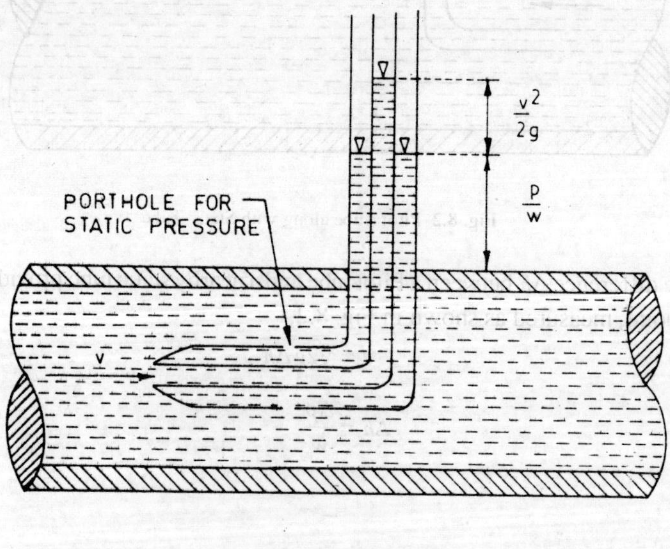

Fig. 8.4 Prandtl Tube

Prandtl tube/Pitot-static tube, as shown in Fig. 8.4, is an improved version of Pitot-tube with a coefficient C equal to unity. It has two tubes concentric with each other. The outer tube has small holes drilled at right angle to the direction of flow and the liquid level in it measures the static head. The inner tube has a hole facing the direction of flow and measures total head. A Pitot static tube combines the measurement of stagnation pressure/total head and static pressure. A differential manometer attached with the Pitot tube measures difference between total pressure and static pressure. This difference in pressure is called dynamic pressure or velocity head.

Demonstration

(i) Take the Pitot tube, inverted U-tube manometer and an empty bucket to a nearby canal/canal distributory in case a channel is not available in hydraulics laboratory.
(ii) Connect the Pitot tube to the manometer through rubber tubings.
(iii) Immerse the Pitot tube in the bucket filled with water and suck air entrapped in the Pitot tube and the manometer till there is no air bubble in the apparatus and water-level is same in both the limbs of the manometer.
(iv) Put the bucket, along with water and the immersed Pitot tube in the canal till the arrangement is below the water surface in the canal.
(v) Take out the bucket taking care that the Pitot tube remains immersed in the canal water.
(vi) Lower the Pitot tube up to the point of measurement of velocity.
(vii) See that the Pitot tube hole is facing the direction of flow.
(viii) With the increase in the velocity of flow in the canal, the manometer reading will increase.

Observations and Calculations

S.No.	Manometer reading		Pressure difference	$v = \sqrt{2g\,\Delta h}$
	h_1	h_2	$\Delta h = h_2 - h_1$	
→ Units				

Discussion of Results

(b) Velocity Distribution in a Pipe

Theory: The velocity of a fluid along a pipe having turbulent flow will vary at different points of the cross-section, its magnitude depending upon the radius of the pipeline. By moving the Pitot tube across the diameter of the pipe, the velocity distribution may be obtained. The variation is shown in Fig. 8.5, the velocity being plotted horizontally on the diameter of the pipe as base. It is found that the velocity is maximum at the centre of the pipe and is minimum at the circumference.

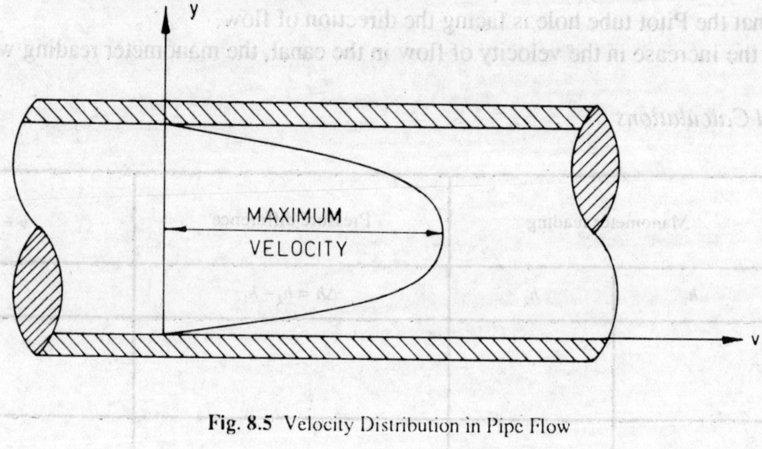

Fig. 8.5 Velocity Distribution in Pipe Flow

Apparatus: It consists of a straight pipeline of about 10 cm diameter and about 5 to 7 m long. Supply to the pipeline is controlled by means of an outlet valve. A pitot-static tube is used to measure the velocity of flow at different points along a vertical section of the pipe, as shown in Fig. 8.6.

Procedure

(i) Note down the reading on the gauge to which pitot-static tube has been fixed, when this tube just touches the lower pipe wall. The reading on the gauge corresponds to $y = d/2$ where d is the diameter of pitot-static tube.

(ii) Open the inlet valve fully and close the outlet valve of the pipeline. Remove the air bubbles, if any, in the manometer tube.

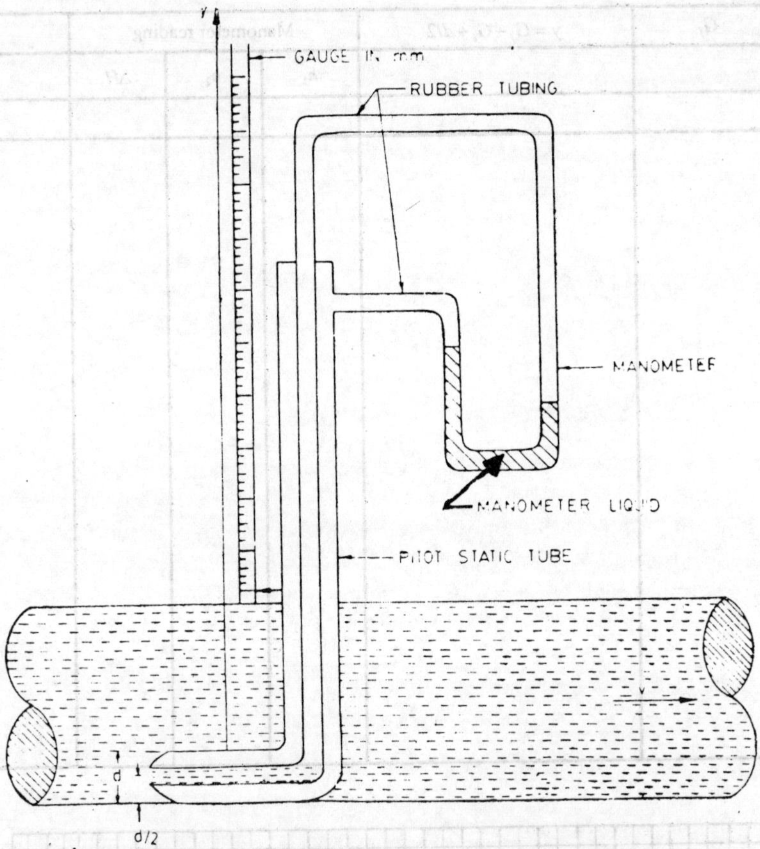

Fig. 8.6 Experimental Set Up for Determining Velocity Distribution in a Pipeline

(iii) Open the outlet valve and allow the flow to take place for some time.

(iv) Note the manometer reading for different positions of pitot-static tube. (Near the wall the interval between successive positions of pitot-static tube may be kept small, say 5 mm, and near the centre this interval may be increased to 10 mm.)

(v) Plot the velocity distribution graph for v vs y.

Observations and Calculations

Diameter of pipe, $D =$

Diameter of pitot-static tube, $d =$

Specific weight of manometer liquid, $w_m =$

Specific weight of water, $w =$

Initial reading of pitot-static tube gauge, $G_i =$

Final reading of pitot-static tube gauge, $G_f =$

S.No.	G_f	$y = G_f - G_i + d/2$	Manometer reading			$v = \sqrt{2g\,\Delta H}$
			h_1	h_2	ΔH	
Units						

Discussion of Results

Precautions

(i) Check that no air bubble is entrapped in Pitot tube or in manometer while taking observations.

(ii) Check that holes in the Pitot tube are open and hence free from dust, etc.

(iii) Pitot tube should not be installed very close to the regulating valve.

EXPERIMENT 9

To study and use current meter in an open channel

Theory: Direct determination of river discharge is usually done by the concurrent measurement of depth and velocity of flow at a suitable section of the stream. Measurement of velocity at a section of the stream is done by pitot-tube, current meter, floats, etc., and depth of flow at section is measured by graduated sounding rods, cables or ultrasonic devices. Current meter, consisting essentially of a rotating element, is used for accurate determination of the stream velocity. Current meters are of two types

(i) Propeller-type current meter

(ii) Cup-type current meter

Both the types of the current meters have the same principle of working. Currentmeter is immersed in the flowing water, which causes the blades/cups of the current meter to rotate. The rotation of these blades/cups, which is a measure of the velocity of flow, is given by sound impulses or electrical impulses. The rotation of blades/cups with respect to time and velocity of flow is calibrated. Corresponding to the number of revolutions made by a current meter in certain time, the velocity of flow is directly read/calculated from the calibration chart.

Description

(i) Propeller-type current meter: It consists of a propeller runner, a revolution counter, a shaft and a tailpiece and can measure velocities in the range of 0.15 m/s to 4 m/s. Propeller is of strictly helical type and is made of one single piece of hard and very resistant aluminium. There is no friction between the axis of the propeller and the electric contact. The current meter is connected to the revolution counter through battery. The contact for transmitting the counting of rotations of the propeller to the electric make and break arrangement inside the casing of the current meter is arranged in a watertight cylindrical chamber. The revolution counter is worked by the electrical impulses sent by current meter and gives number of revolutions made by the propeller of the current meter. A heavy circular weight is attached at the bottom and the entire instrument is then fixed to a vertical rod or a cable.

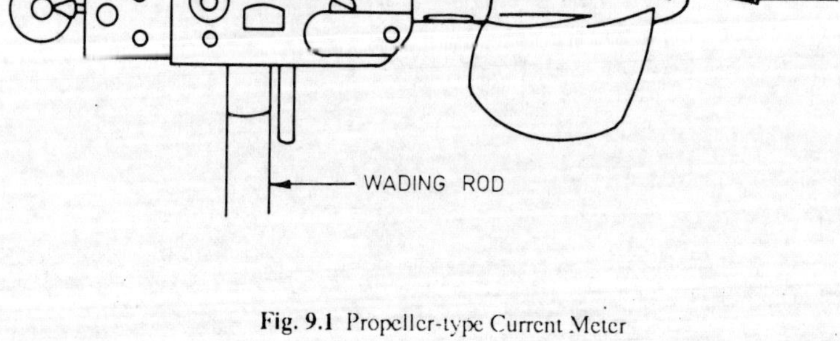

Fig. 9.1 Propeller-type Current Meter

(*ii*) *Cup-type current meter:* This is practically the same as the propeller-type except that instead of helical propeller there are cups attached to a vertical spindle. The cups rotate in a horizontal plane, thereby rotating the vertical shaft. To the vertical spindle is attached the cam which records the revolutions. The dead weight to be attached at the bottom is of the shape of a fish and has fins and is called fish weight.

This type of current meter is capable of measuring velocities from 0.15 m/s to 3.4 m/s; though miniature current meter, also called pigmy current meter, is capable of measuring velocities up to 1 m/s.

Both these types of current meters can measure velocity of flow at any required depth of a given channel.

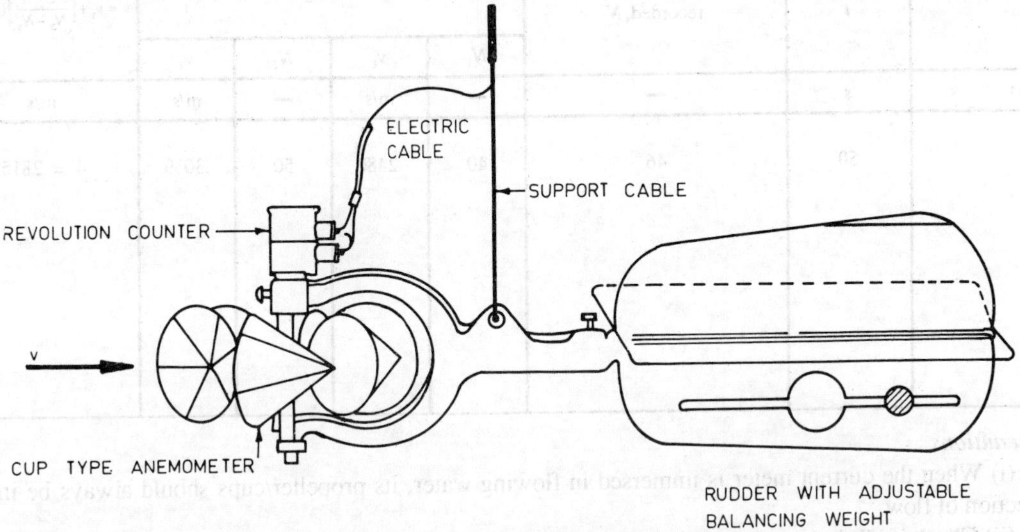

Fig. 9.2 Cup-type Current Meter

Demonstration

(i) Let us assemble the current meter (propeller-type or cup-type) available in hydraulics laboratory.

(ii) The current meter is lowered to desired depth in an open channel carrying water in hydraulics Laboratory or in a nearby canal/canal distributory.

(iii) See that cup/propeller side of the current meter is in the direction of flow of water.

(iv) We find that the propeller/cups is/are rotating.

(v) Now as we switch on the revolution counter, it starts counting the revolutions of propeller/cups.

(vi) In case a headphone is supplied with the current meter, it is put on the ear and a tick sound is heard corresponding to each revolution of the propeller/cups of the current meter.

(vii) If N is the number of revolutions made by the propeller/cups of the current meter in a known time, t, then corresponding to this value of N and t, the velocity of flow can be determined with the help of the rating table of the current meter. (Consult the rating table given on page 65.)

Observations and Calculations

Velocity of flow corresponding to N_1 revolutions of current meter made in t seconds = v_1
Velocity of flow corresponding to N_2 revolutions of current meter made in t seconds = v_2
The values of N_1, N_2, v_1 and v_2 corresponding to time t seconds, are read from the rating table of the current meter, so that velocity of flow, v, corresponding to N revolutions of current meter made in t seconds could be calculated as shown below

S.No.	Time t	No. of revolution recorded, N	From rating table				$v = v_1 + \left(\frac{v_2 - v_1}{N_2 - N_1}\right)(N - N_1)$
			N_1	v_1	N_2	v_2	
Units	s	—	—	m/s	—	m/s	m/s
1.	50	46	40	.2480	50	.3039	$V = .2815$
2.							
3.							
4.							
5.							

Precautions

(i) When the current meter is immersed in flowing water, its propeller/cups should always be in the direction of flow.

(ii) Check if all the nuts are watertight.

(iii) Electrical connections should not be loose.

(iv) Friction between the moving parts should be minimum.

Central Water and Power Research Station, Khadakwasla, Poona
Current Meter Rating Table

Time (s)	5	8	10	15	20	30	40	50	60	70	80	90	100	150	200	250	Time (s)
								No. of Revolutions V-Velocity, m/s									
40	—	—	.0943	.1293	.1642	.2340	.3039	.3738	.4436	.5135	.5834	.6532	.7231	1.0714	1.4186	1.7659	40
41	—	—	.0926	.1267	.1608	.2289	.2971	.3652	.4334	.5016	.5697	.6379	.7060	1.0460	1.3848	1.7236	41
42	—	—	.0910	.1243	.1575	.2241	.2906	.3571	.4237	.4902	.5567	.6233	.6898	1.0218	1.3525	1.6832	42
43	—	—	—	.1219	.1544	.2194	.2844	.3494	.4144	.4794	.5444	.6093	.6743	.9987	1.3217	1.6448	43
44	—	—	—	.1197	.1515	.2150	.2785	.3420	.4055	.4690	.5325	.5961	.6596	.9767	1.2924	1.6081	44
45	—	—	—	.1176	.1487	.2108	.2729	.3350	.3971	.4592	.5213	.5834	.6455	.9556	1.2643	1.5730	45
46	—	—	—	.1156	.1460	.2067	.2675	.3282	.3890	.4497	.5105	.5712	.6320	.9355	1.2375	1.5394	46
47	—	—	—	.1136	.1434	.2028	.2623	.3217	.3812	.4407	.5001	.5596	.6190	.9162	1.2118	1.5073	47
48	—	—	—	.1118	.1409	.1991	.2573	.3155	.3738	.4320	.4902	.5484	.6066	.8977	1.1871	1.4765	48
49	—	—	—	.1100	.1385	.1955	.2526	.3096	.3666	.4237	.4807	.5377	.5948	.8799	1.1635	1.4470	49
50	—	—	—	.1083	.1362	.1921	.2480	.3039	.3598	.4157	.4716	.5275	.5834	.8628	1.1408	1.4185	50
51	—	—	—	.1066	.1340	.1888	.2436	.2984	.3532	.4080	.4628	.5176	.5724	.8464	1.1190	1.3914	51
52	—	—	—	.1051	.1319	.1857	.2394	.2932	.3469	.4006	.4544	.5081	.5619	.8306	1.0981	1.3652	52
53	—	—	—	.1035	.1299	.1826	.2354	.2881	.3408	.3935	.4463	.4990	.5517	.8153	1.0779	1.3400	53
54	—	—	—	.1021	.1280	.1797	.2315	.2832	.3350	.3867	.4385	.4902	.5420	.8007	1.0585	1.3158	54
55	—	—	—	.1007	.1261	.1769	.2277	.2785	.3293	.3801	.4309	.4817	.5325	.7866	1.0398	1.2924	55
56	—	—	—	.0993	.1243	.1742	.2241	.2740	.3239	.3738	.4237	.4736	.5235	.7730	1.0218	1.2698	56
57	—	—	—	.0980	.1225	.1715	.2206	.2696	.3186	.3676	.4167	.4657	.5147	.7598	1.0044	1.2481	57
58	—	—	—	.0967	.1208	.1690	.2172	.2654	.3135	.3617	.4099	.4581	.5063	.7472	.9876	1.2271	58
59	—	—	—	.0955	.1192	.1666	.2139	.2613	.3086	.3560	.4034	.4507	.4981	.7349	.9713	1.2068	59
60	—	—	—	.0943	.1176	.1642	.2108	.2573	.3039	.3505	.3971	.4436	.4902	.7231	.9556	1.1871	60
61	—	—	—	.0932	.1161	.1619	.2077	.2535	.2993	.3451	.3909	.4368	.4826	.7116	.9404	1.1682	61
62	—	—	—	.0921	.1146	.1597	.2047	.2498	.2949	.3400	.3850	.4301	.4752	.7005	.9258	1.1498	62
63	—	—	—	.0910	.1132	.1575	.2019	.2462	.2906	.3350	.3793	.4237	.4680	.6898	.9115	1.1320	63
64	—	—	—	.0900	.1118	.1554	.1991	.2428	.2864	.3301	.3738	.4174	.4611	.6794	.8977	1.1148	64
65	—	—	—	—	.1104	.1534	.1964	.2394	.2824	.3254	.3684	.4114	.4544	.6693	.8843	1.0981	65
66	—	—	—	—	.1091	.1515	.1938	.2362	.2785	.3208	.3632	.4055	.4479	.6596	.8713	1.0819	66
67	—	—	—	—	.1079	.1496	.1913	.2330	.2747	.3164	.3581	.3998	.4415	.6501	.8586	1.0662	67
68	—	—	—	—	.1066	.1477	.1888	.2299	.2710	.3121	.3532	.3943	.4354	.6409	.8464	1.0510	68
69	—	—	—	—	.1055	.1460	.1865	.2270	.2675	.3080	.3485	.3890	.4295	.6320	.8344	1.0362	69
70	—	—	—	—	.1043	.1442	.1841	.2241	.2640	.3039	.3438	.3837	.4237	.6233	.8229	1.0218	70

(Contd.)

S.No. of Current Meter 8727
Meter 3 Type 108 Semitron
Date of Rating 5.12.74 (Pigmy)

Spin test 25 S
Threshold Velocity .0900, Normal/High
Rating Range 0.1 To 2.0 M/S,
Dispersion for Velocity
Is beyond Stipulated one

Rating Equation with Rigid Rod Suspension
For $R \leq 3.1375$ $V = .0245 + .2794\,R$
For $R \geq 3.1375$ $V = .0296 + .2778\,R$

Advised to use Current Meter for Velocities .1800 to 2 m/s

V-Velocity, m/s R-Revolutions/s

Date of Statistical Interest

$V_t = .120$	.207	.344	.437	.675	.864	1.284	1.481	1.864
$R = .312$	.686	1.139	1.494	2.331	2.996	4.502	5.253	6.594
Ratio $V_t/V = 1.0776$	.9606	1.0054	.9899	.9991	1.0034	1.0032	.9947	1.0015

V_t-Trolly Velocity, m/s V-Velocity from Rating Equation, m/s

M	AI	BI	SDI	PDI	A2	B2	SD2	PD2	VC	RC	F	N
5	.0245	.2794	.0067	1.5215	.0296	.2778	.0057	.4638	.9012	3.1375	5	13

3 Jan 1975

Sd/-xxxxxxxxxxxxx
Tank Superintendent,
Ship Hydrodynamics
C.W. and P.R.S. Poona - 441024.

EXPERIMENT 10

To establish the relationship between surface and mean velocity of flow in a channel by surface float.

Theory: While current meter should be used as far as possible for the measurement of velocity of flow in open channels; under certain circumstances such as

(i) Velocities of flow being beyond the calibrated range of current meter,

(ii) the flow being such that there is a danger of losing current meter,

(iii) there being lack of facilities for use of current meter, it may not be possible or advisable to use a current meter and in such circumstances surface floats and velocity rods may have to be used. While the surface floats give only surface velocity, velocity rods directly give the mean velocity to a fair degree of accuracy. However, the use of velocity rods is restricted to sites where the bed of the stream is uniform and free of large humps, that is for measurement of velocities in regular canals. Surface floats give the surface velocity, and *the mean velocity of flow at a section is calculated by multiplying the observed surface velocity by 0.8.* Though in certain cases the mean velocity may be as low as 0.65 and as high as 0.81 of surface velocity, yet for all practical purposes the coefficient may be taken as 0.8 which is suitable for average conditions.

Description

(a) Floats: Depending upon the nature and condition of use, surface floats are generally made of wood or other suitable materials. Wooden floats are meant to be used in rivers under moderate velocity conditions with less turbulence, while metallic floats are meant for large turbulent rivers. Bottle floats weighted with cement mortar can also be used in relatively smooth flow conditions. Generally the wooden floats consist of

(a) Circular disc,

(b) Cylindrical stem and hooks, and

(c) Flag

The circular disc is either streamlined at the edges or cylindrical in shape as shown in Fig. 10.1.

(b) Velocity Rods: They consist of vertical wooden rods of varied lengths, 2.5 to 5 cm in diameter, weighted at the bottom to keep them vertical. Generally these rods are submerged up to 0.95 of the depth of water, which is a variable quantity, it is difficult to recommend a fixed length for these rods. Velocity rods are broadly classified into two types:

(i) Velocity rod, rigid (single tube) and
(ii) Telescopic (adjustable) velocity rod.

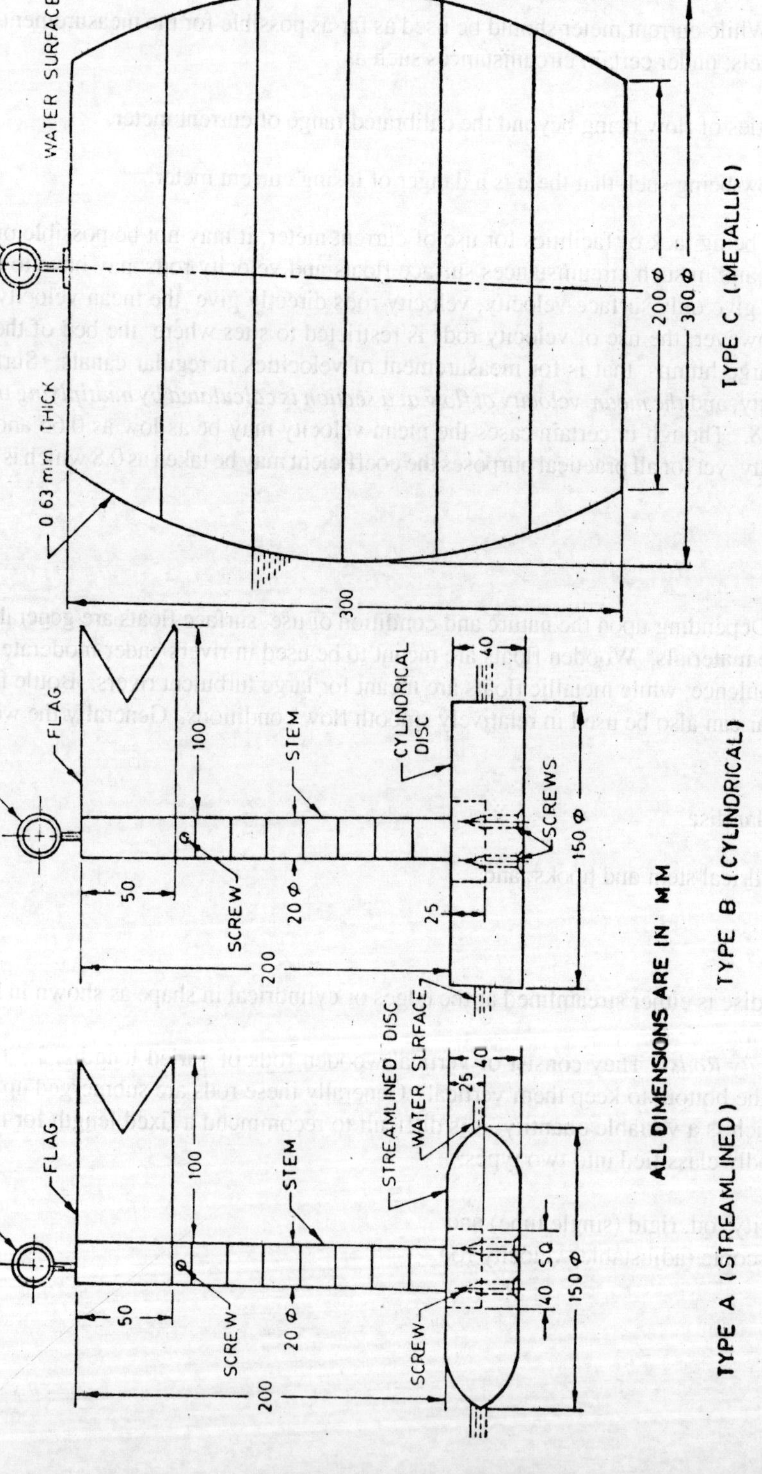

Fig. 10.1 Details of Surface Floats

(i) Rigid Velocity Rod (Single Tube): This type of rod is used when the depths are uniform and do not vary from vertical to vertical. These may be made of wood or galvanised steel sheet as shown in Fig. 10.2.

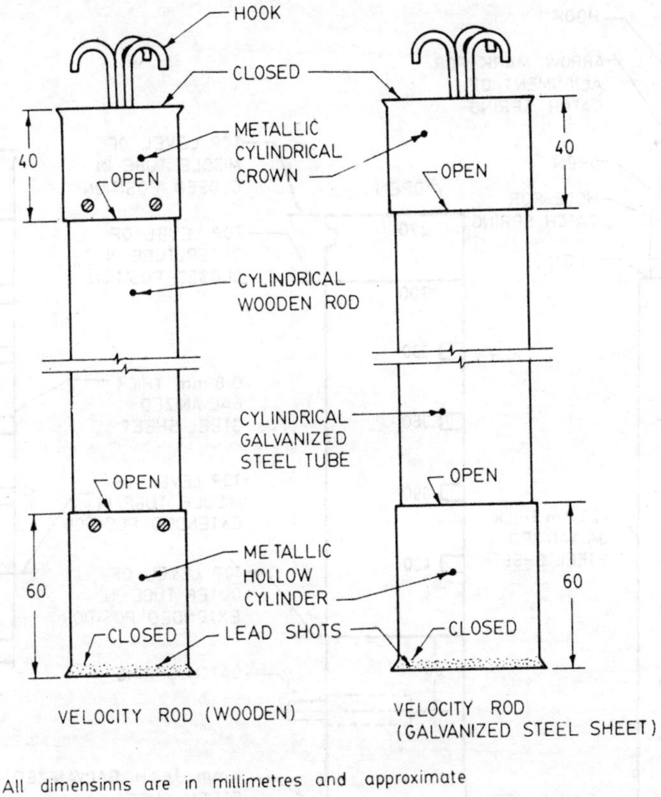

Fig. 10.2 Velocity Rods

(ii) Telescopic Velocity Rod (Adjustable): This type of rod is useful when rods of different lengths are required as their lengths are adjustable. The life of these rods is less than that of the rigid rods. Because of constant alteration in length, holes get bigger and lead to drowning of rods. These rods are manufactured to read from 300 to 600 mm, 450 to 750 mm, 600 to 1200 mm, 1 to 2 m, 2 to 3 m, and 3 to 5 m. It consists of three metallic hollow tubes, one working inside the other and each closed at the bottom as shown in Fig. 10.3.

Demonstration

(i) Let us go to the nearby canal/canal distributory.

(ii) Choose two stations 100 to 150 m apart along the length of the canal. This distance is also called the run.

(iii) Measure the distance between the two marked stations.

(iv) Allow the float/velocity rod to float from one station to another station and note down the time taken to cover this run.

(v) Repeat step (iv) to take two more readings for the same run (or three floats can be made to float simultaneously between two stations and their respective time noted).

(vi) Repeat steps (iv) and (v) for two more runs.

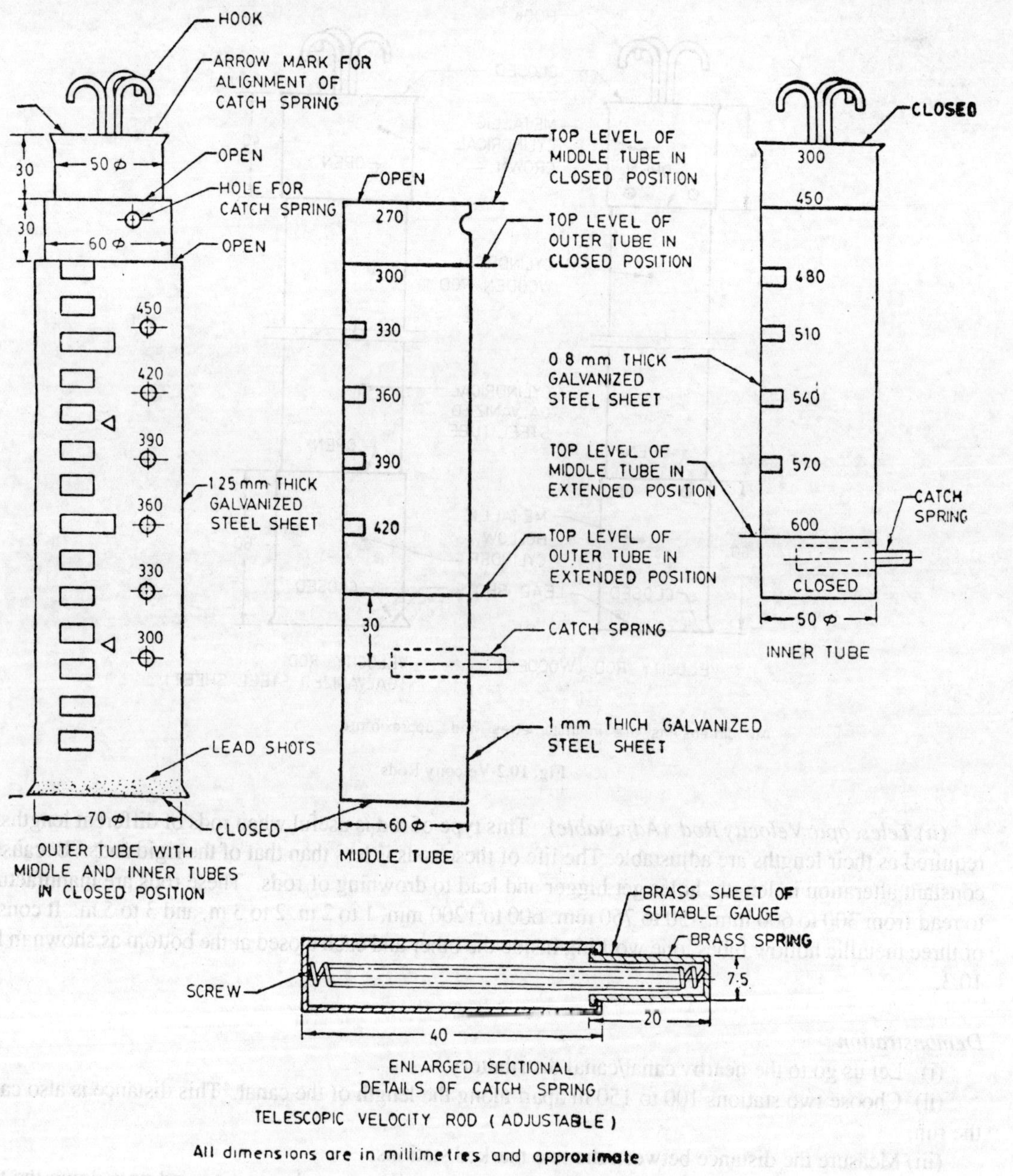

TELESCOPIC VELOCITY ROD (ADJUSTABLE)

All dimensions are in millimetres and approximate

Fig. 10.3 Telescopic Velocity Rod

Observations and Calculations

S.No.	Measurement by	Run No.	Distance between two stations	Time	Surface velocity	Mean velocity
Units					Average	Average
	Float	I II III				
	Velocity rod	I II III				

Since surface velocity of flow = K times mean velocity of flow

$$K = \frac{\text{surface velocity}}{\text{mean velocity}}$$

Hence the relationship between surface velocity and mean velocity of flow in a channel is

$$\text{Surface velocity} = \ldots \times \text{mean velocity}$$

Discussion of Results

Precautions

(i) Care should be taken to see that weeds at the bottom of the stream do not interfere with the use of a velocity rod. If possible a section of stream which is free from weeds should be chosen.

(ii) Since the observations with floats are liable to be influenced by wind, waves and eddies, care should be taken to make measurements along long stretch of river on a calm/quite day.

(iii) When the time of run differs considerably from the readings of the successive or adjacent floats, the particular reading may be discarded.

(iv) The submerged length of the rod should be 0.94 times the depth of the stream, subject to the condition that it shall float 7 to 15 cm short of the bottom of the stream.

EXPERIMENT 11

To find the coefficient of discharge C_d for a rectangular, triangular and trapezoidal sharp-edged notch.

Theory: In hydraulics engineering, notches are commonly used to regulate flow in rivers and other open channels. The relationship between the water-level upstream of the notch and the discharge over it is, generally, known so that the discharge at any time may be found by observing the upstream water-level. Notches usually have sharp edges so that the water springs clear of the plate as it passes through the notch.

With the usual notations, the equation for discharge past a sharp-edged notch can be derived in the following form:

$$Q = \frac{2}{3}\sqrt{2g}\, b h^{3/2} \qquad \text{(i)}$$

for the rectangular notch;

$$Q = \frac{8}{15}\sqrt{2g}\tan\frac{\theta}{2} h^{5/2} \qquad \text{(ii)}$$

for the triangular notch; and

$$Q = \frac{2}{3}\sqrt{2g}\, b h^{3/2} + \frac{8}{15}\sqrt{2g}\tan\frac{\theta}{2} h^{5/2} \qquad \text{(iii)}$$

for the trapezoidal notch

where h, b, θ are as shown in Fig. 11.1.

The preceding equations are based upon the assumptions listed as follows:

(a) The height of the nappe at the plane of the notch opening is h.

(b) The velocity of flow is normal to the plane of the notch opening at all points.

(c) The velocity in the approach channel is negligible.

(d) There is no contraction of the stream as it passes through the notch.

In practice, however, none of these assumptions are satisfied. It is, therefore, customary to rewrite the equations in the form

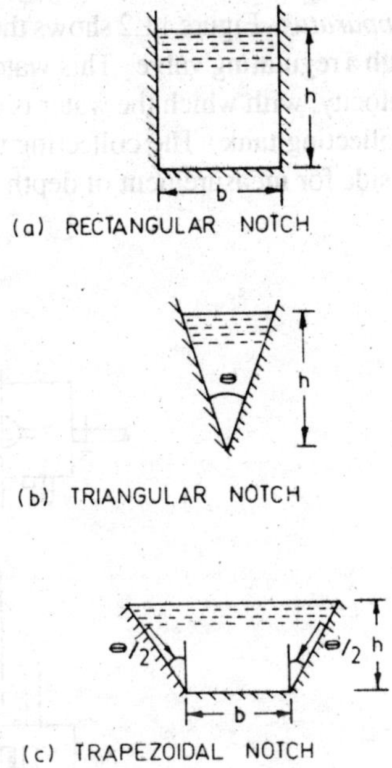

Fig. 11.1 Various Types of Notches

$$Q = C_d \frac{2}{3}\sqrt{2g}\, b h^{3/2} \qquad \text{(iv)}$$

for the rectangular notch;

$$Q = C_d \frac{8}{15}\sqrt{2g}\tan\frac{\theta}{2} h^{5/2} \qquad \text{(v)}$$

for the triangular notch; and

$$Q = C_d \frac{2}{3}\sqrt{2g}\, b h^{3/2} + C_d \frac{8}{15}\sqrt{2g}\tan\frac{\theta}{2} h^{5/2} \qquad \text{(vi)}$$

for the trapezoidal notch,

where C_d is a coefficient of discharge of the notch. It is a non-dimensional number and the dependable way of its determination is by experimentation. A convenient way of finding C_d and the exponent of h in either of these expressions is as follows. Either of the Eqs (iv), (v) and (vi) may be written in the form:

$$Q = K(h)^n \quad \text{(vii)}$$

$$\log Q = \log K + n \log h \quad \text{(viii)}$$

If experimental results are plotted on a graph having log h as abscissa and log Q as ordinate, then provided that K and n are constant over the range of the results, they will lie on a straight line having slope n and intercept log K on the axis of log Q, as indicated in Fig. 11.4.

Apparatus: Figure 11.2 shows the arrangement in which water from overhead tank is fed to the apparatus through a regulating valve. This water is made to pass through the holes of the slotted plates, so as to reduce the velocity; with which the water is approaching the notch. Water, which flows over the notch, is collected in a collecting tank. The collecting tank has an outlet on the side of the tank and a piezometer tube on the other side for measurement of depth of water in the collecting tank.

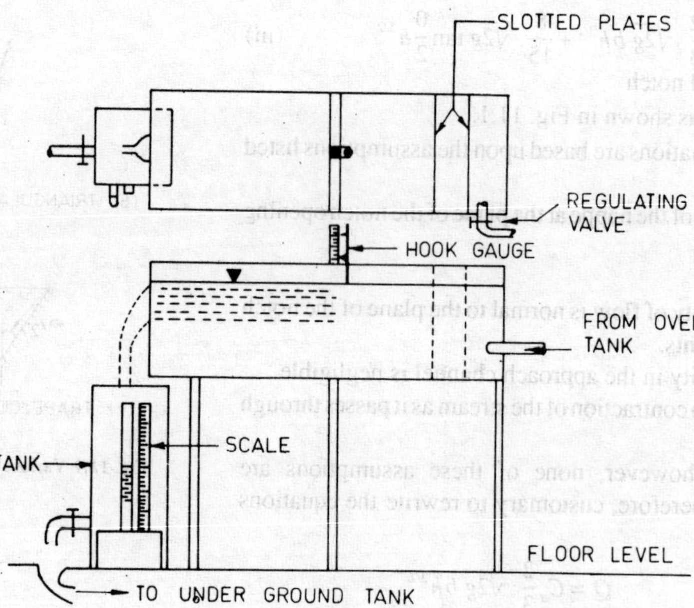

Fig. 11.2 Notch Apparatus

The apparatus carries a hookgauge, which consists of a sharply-pointed hook connected to the end of vertical rod attached to a vertical scale. The hook is best set in the surface by viewing it from top so that its image may be seen in the water surface. The hook is then slowly raised until the hook and its image just coincide as the hook touches the water surface.

Experimental Procedure

(i) Record the width of the rectangular notch; the angle of V-notch (best found by measuring the depth and width of the V in case of triangular notch) and the width of the notch and the side slope in case of trapezcidal notch.

(ii) Level the apparatus and establish the zero of the hook gauge reading corresponding to the level of the crest of the notch.

To do this, admit water from the overhead tank to the apparatus until the level is approximately correct, and then carefully bale out or in, using a small beaker, until correct level is reached. For the rectangular/trapezoidal notch, this can be checked as illustrated in Fig. 11.3 (a) by placing a steel rule on the crest. For the triangular notch, the reflection of the V in the surface serves to indicate whether the level is correct or not as illustrated in Fig. 11.3 (b).

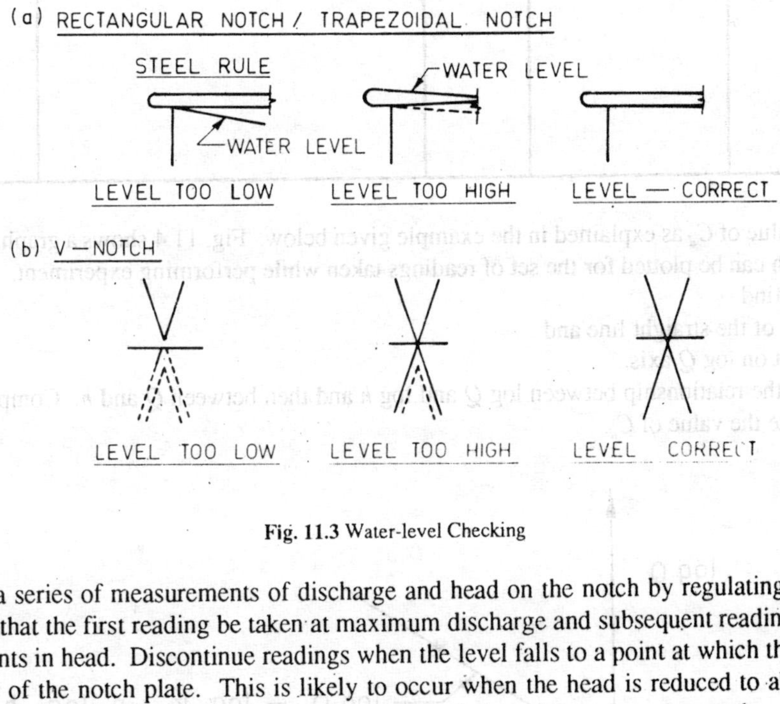

Fig. 11.3 Water-level Checking

(iii) Take a series of measurements of discharge and head on the notch by regulating the flow. It is recommended that the first reading be taken at maximum discharge and subsequent readings with roughly equal decrements in head. Discontinue readings when the level falls to a point at which the stream ceases to spring clear of the notch plate. This is likely to occur when the head is reduced to about 1 cm for a rectangular notch 1.5 cm for a trapezoidal notch and about 2.5 cm for a triangular notch.

About 8 different discharges for each notch should be sufficient.

Observations and Calculations

(*i*) *Rectangular Notch*

Width of notch, b =

No.	Gauge reading h	Quantity of water collected, V_o	Time t	Discharge Q	Calculation of C_d
Unit →					
1.					
2.					
3.					
4.					
5.					
6.					
7.					
8.					

Calculate the value of C_d as explained in the example given below. Fig. 11.4 shows a graph between log Q and log h, which can be plotted for the set of readings taken while performing experiment.

From Fig. 11.4 find

(i) Slope n, of the straight line and

(ii) Intercept on log Q axis.

From this, find the relationship between log Q and log h and then between Q and h. Comparing this with Eq. (iv) calculate the value of C_d.

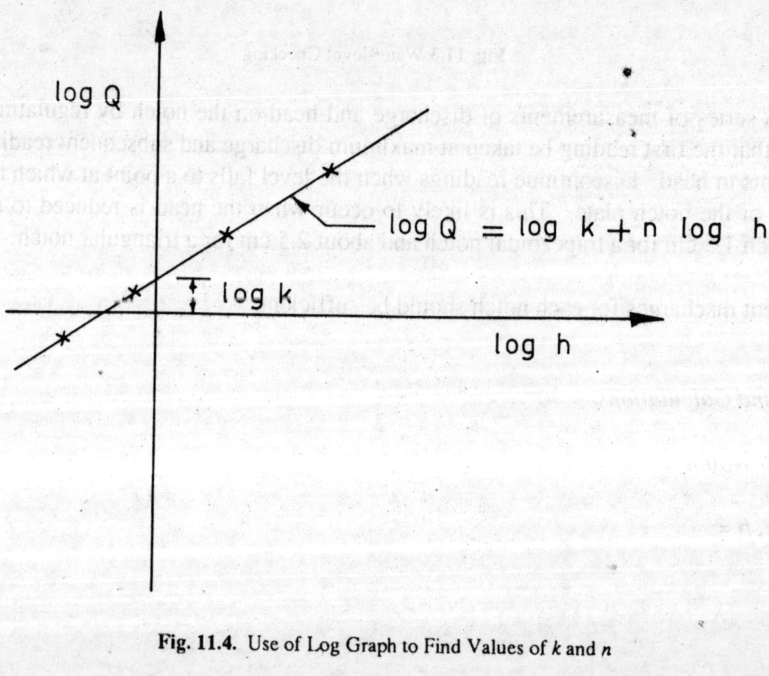

Fig. 11.4. Use of Log Graph to Find Values of k and n

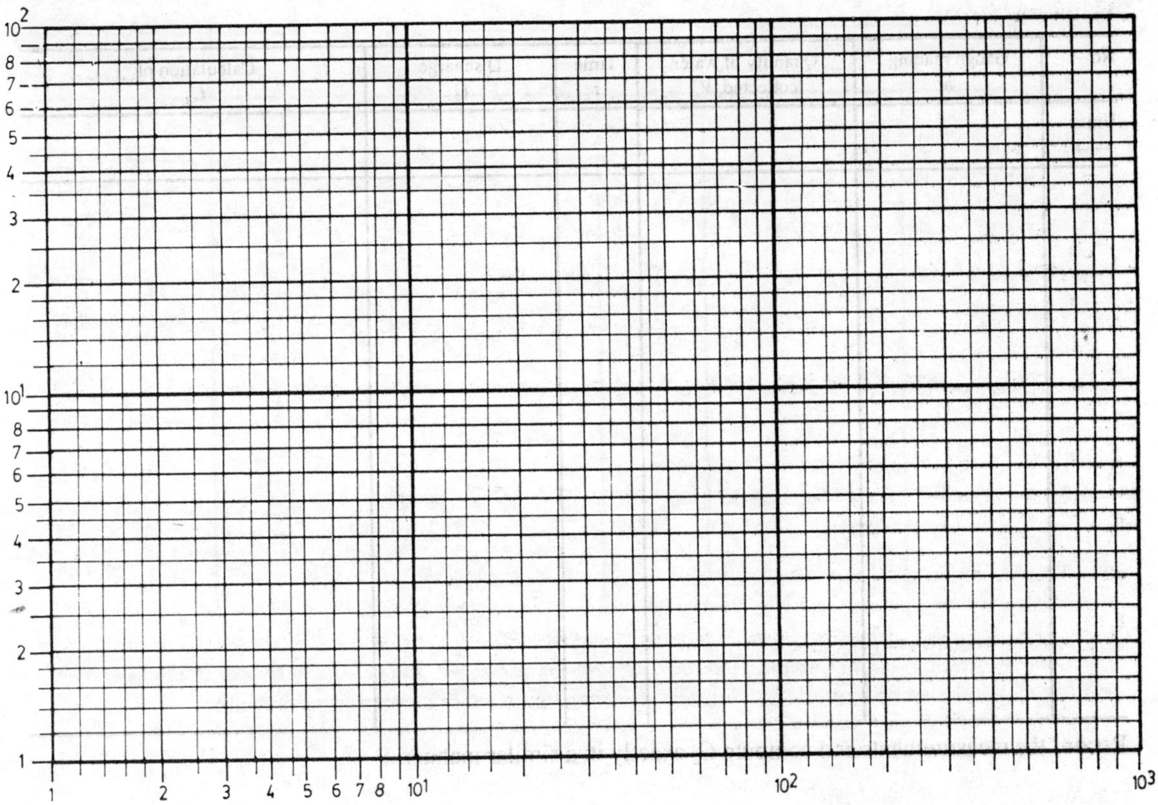

For example, in Fig. 11.4, slope $n = 1.50$
Intercept on $\log Q$ axis = 2.248
The relationship between $\log Q$ and $\log h$ is thus, $\text{Log } Q = 2.248 + 1.50 \log h$
so that the relationship between Q and h is, $Q = 177.0\, h^{3/2}$

Comparing this with $Q = C_d \frac{2}{3}\sqrt{2g}\, b h^{3/2}$

it is noted that the exponent of h is the same in both and that C_d is given by

$$C_d = \frac{177.0}{\frac{2}{3}\sqrt{2g}\, b} \doteq 0.596 \text{ or } 0.60$$

(ii) Triangular Notch
Width across the top of V-notch, b =
Depth of V-notch, d =

$$\tan\frac{\theta}{2},\ b/2d =$$

No.	Gauge reading h	Quantity of water collected, V_o	Time t	Discharge Q	Calculation of C_d
Units →					
1.					
2.					
3.					
4.					
5.					
6.					
7.					
8.					

Record the measurements and compute C_d exactly in a similar manner.

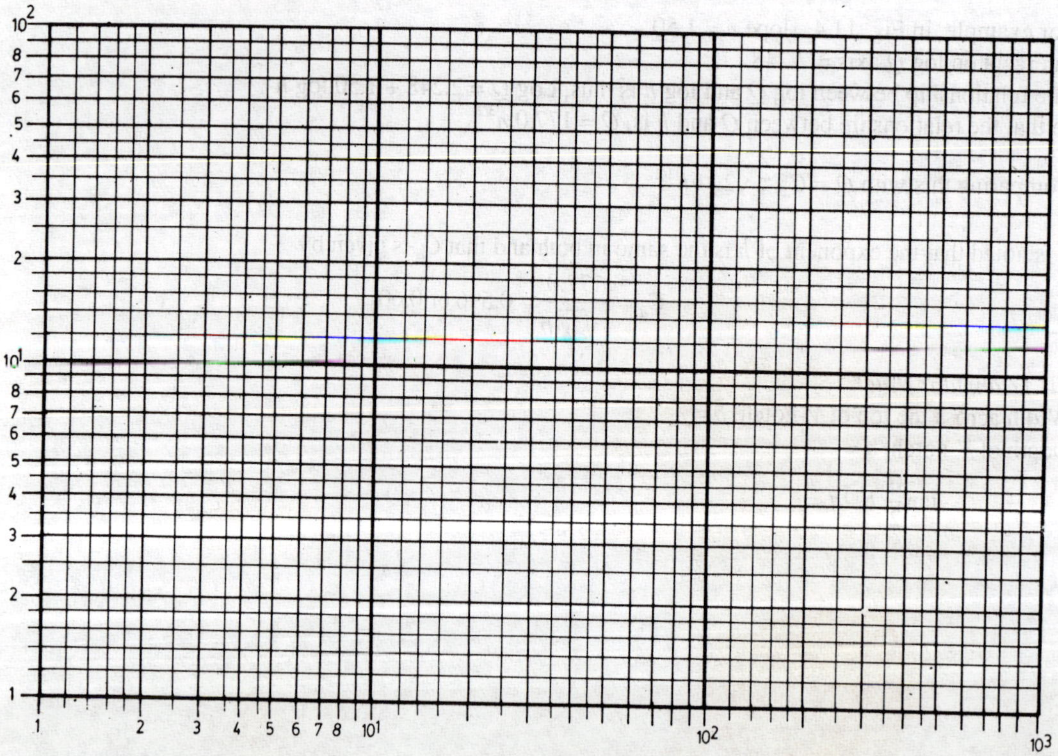

(iii) Trapezoidal Notch

Bottom width of the notch, b = Top width of the notch, B =

Depth of the notch, $d = \tan\frac{\theta}{2}, \frac{(B-b)}{2d} =$

No.	Gauge reading h	Quantity of water collected, V_0	Time t	Discharge Q	Calculation of C_d
Units →					
1.					
2.					
3.					
4.					
5.					
6.					
7.					
8.					

Record the measurements and compute C_d exactly in a similar manner.

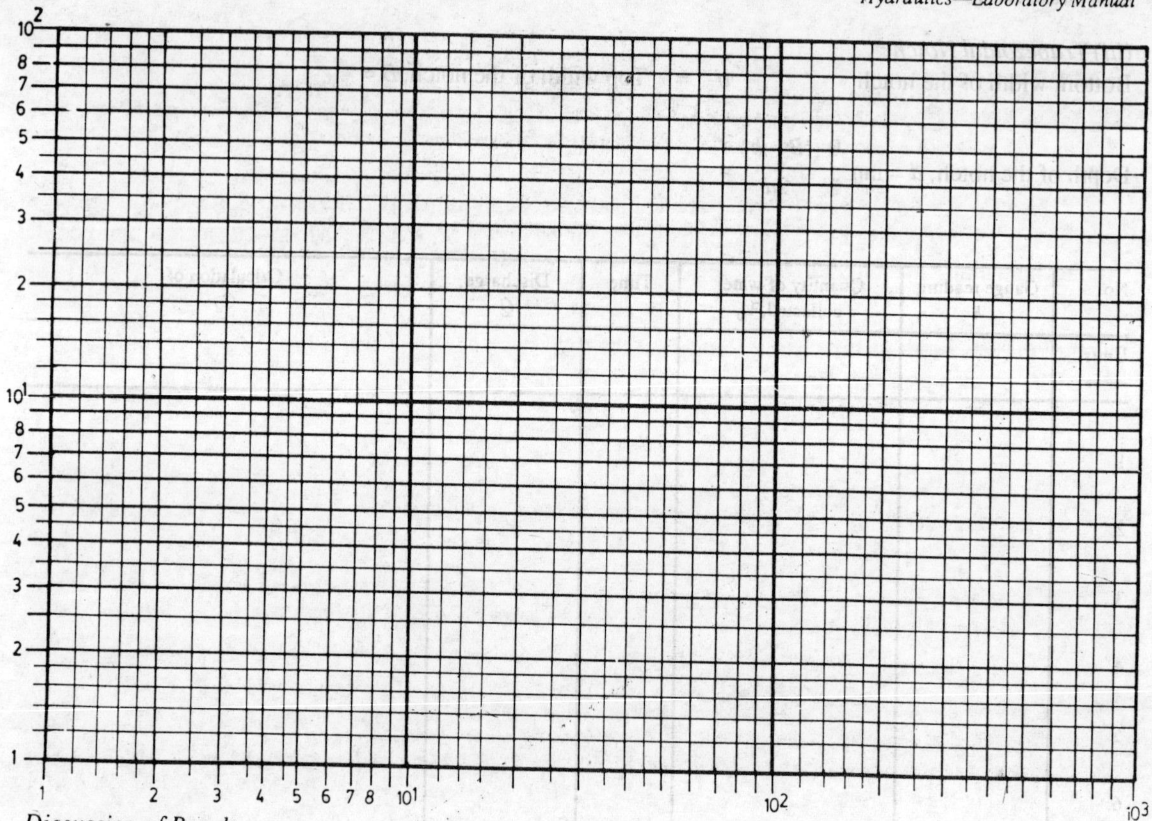

Discussion of Results

Questions for Discussion

(i) How would you interpret results, which, when plotted logarithmically, as in Fig. 11.4, fall on a line which is not straight, but slightly curved.

Precautions

(i) As the correct discharge measurements are very important for this experiment, there should be no leakage at any of the regulating valves.

(ii) The width of notch or the angle of the V-notch should be carefully recorded.

(iii) The apparatus should be levelled.

(iv) It is very important to establish the zero of the hook gauge reading corresponding to the level of the crest of the notch. Therefore, as explained in Fig. 11.3, a check that the crest of the notch just lies in the water surface must be made before doing the experiment.

(v) For measurement of h, the hook gauge reading should be taken a little distance away from the crest of the notch.

EXPERIMENT 12

To find the value of coefficient of discharge for broad-crested weir.

Theory: A weir is said to be broad-crested if the stream lines are parallel to the surface of the crest. It should have substantial crest length in the direction of flow. So long as the flow is critical and parallel somewhere along the length of weir, the discharge is given by

$$Q = CbH^{3/2} \tag{i}$$

where b is the width of the channel, and H is the total head acting over the broad-crested weir measured upstream as shown in Fig. 12.1.

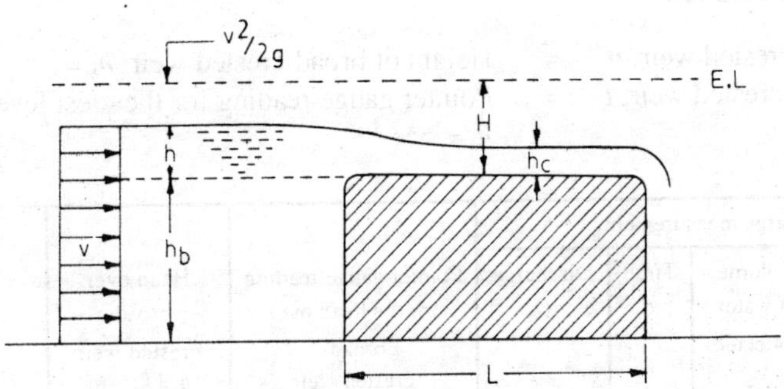

Fig. 12.1 Broad Crested Weir

If, however, the velocity v of the liquid upstream is very small, then $\frac{v^2}{2g}$ is neglected and Eq. (i) is reduced to

$$Q = Cbh^{3/2} \tag{ii}$$

where h is the depth of flow above the crest of broad-crested weir up to free surface level in the approach channel.

Apparatus: It consists of a rectangular channel, say 5 m long, 0.4 m wide and 0.7 m deep. A broad-crested weir having a length of 0.4 m and height about 0.15 m is constructed sufficiently upstream of the outlet end of channel. Top rails for the movement of the pointer gauge are provided along the side walls of the channel. If the supply is made through a supply pipe, a sluice valve is provided at the outlet end of the supply pipe. A pointer gauge is provided for measuring the depth of flow.

Experimental Procedure

(i) Take the pointer gauge reading when the pointer is just touching the crest of the broad-crested weir with no flow in the channel. Also measure the width b and length L of the broad-crested weir.

(ii) Open the sluice gate fully and locate a section in the channel upstream the weir where the water surface has no curvature even at the maximum discharge and adjust the pointer gauge at this position.

(iii) Allow the flow to become steady.

(iv) Measure the discharge of the weir with the help of a collecting tank and take the corresponding water surface level reading with the help of pointer gauge.

(v) Change the flow and repeat steps (iii) and (iv). Take at least six readings.

(vi) Plot h vs Q (with h along X-axis) on a log-log graph paper and fit in a straight line to join maximum number of points.

Observations and Calculations

Width of broad-crested weir, b = Height of broad-crested weir, h_b =

Length of broad-crested weir, L = Pointer gauge reading for the crest level of broad-crested weir h_i =

S.No.	Discharge measurement: Increase in depth of water tank, d	Volume of water collected V_0	Time t	Discharge Q	Pointer gauge reading for the head over broad-crested weir h_f	Head over broad-crested weir $h = hf - hi$	Calculation of C_d
Units →							

Since

$$Q = C b h^{3/2}$$

i.e.,

$$Q = K h^n \text{ where } n = \frac{3}{2} \text{ and } K = C b$$

$$\log Q = \log K + n \log h$$

$$\log Q = n \log h + \log K$$

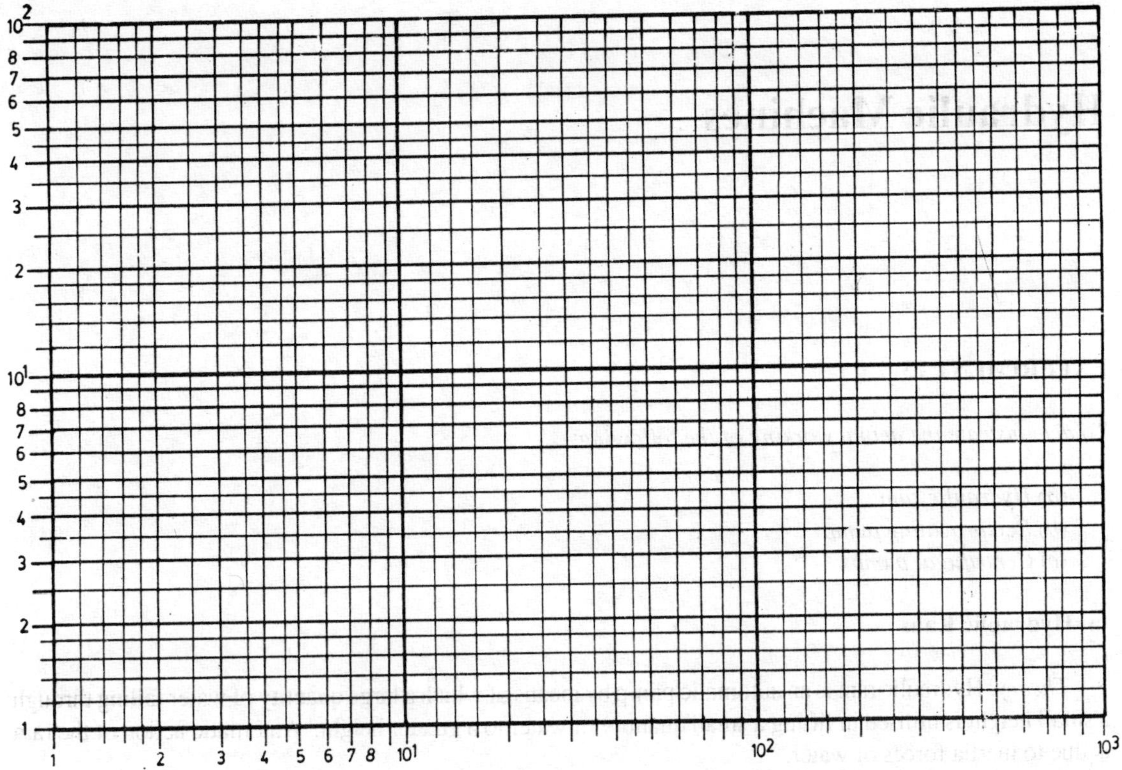

Graph between log Q and log h will be a straight line whose slope will be equal to 3/2 and its intercept on Y-axis is given by log K. Measuring slope of the line and intercept made by it on Y-axis, the values of n and K can be determined.

$$\text{Since } Cb = K \quad \therefore \; C = \frac{K}{b}$$

Discussion of Results

Precautions

(i) Pointer gauge reading should be measured accurately.

(ii) Check that the streamlines are parallel to the top of the broad-crested weir before making observations of Q and h.

Hydraulic Machines

EXPERIMENT 13

To demonstrate the actual working of the following:

(*a*) *Hydraulic ram*
(*b*) *Reciprocating pump*
(*c*) *Centrifugal pumps*

(a) Hydraulic Ram

Theory: Hydraulic ram is an automatic pump by means of which a large quantity of water falling through a small height is utilised in lifting a small quantity of water to a greater height. Automatic action of the ram is due to inertia forces of water.

Apparatus: Fig. 13.1 shows the details of hydraulic ram. It consists of a supply reservoir connected to a ram body through the drive pipe. The ram body consists of a chamber connected to drive pipe and to delivery pipe through an air vessel. Water from the chamber can either pass through the waste water valve C to the waste or through delivery valve F to the air vessel.

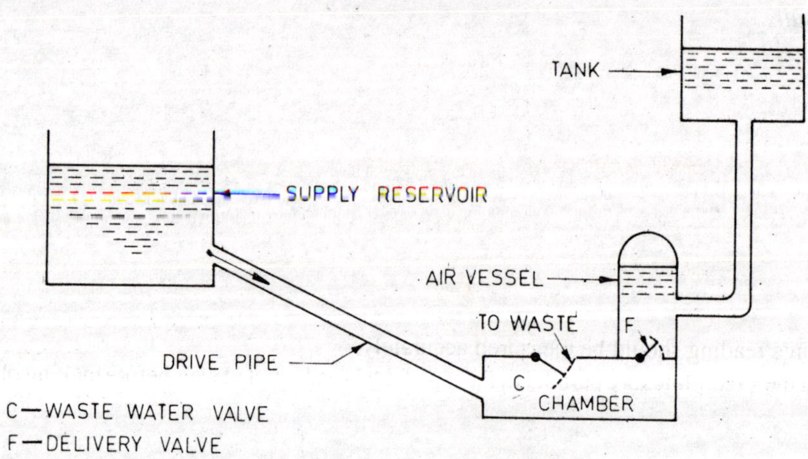

Fig. 13.1 Hydraulic Ram

Demonstration

(i) Water from supply reservoir flows down the drive pipe into the chamber.

(ii) Waste water valve C is opened and the water flows through it to the waste.

(iii) As the speed of water in drive pipe increases, the dynamic pressure on the valve C increases, until it is ultimately greater than the weight of the valve lid, so that the valve will suddenly close.

(iv) Closing of valve C brings the water in drive pipe suddenly to rest, causing an increase in pressure in the chamber.

(v) This increase in pressure lifts the delivery valve F and some of the water flows into air vessel and compresses the air in the vessel.

(vi) This increased air pressure forces the water into the tank.

(vii) As the momentum of water in the chamber is destroyed, the valve F closed and valve C opens, causing the flow from supply reservoir to recommence and the cycle is repeated.

(b) Reciprocating Pump

Theory: A reciprocating pump is a device, driven by power from an external source, which when interposed in a pipe transfers mechanical energy to the liquid flowing thorough the pipe.

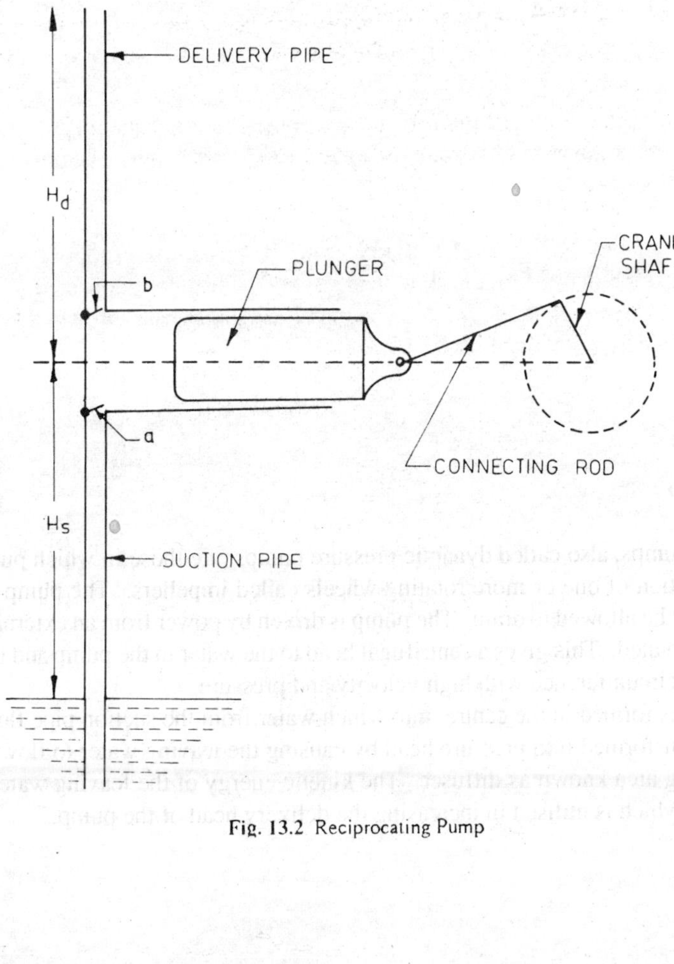

Fig. 13.2 Reciprocating Pump

Apparatus: It consists of a cylinder in which a piston or plunger is moved backward and forward. Cylinder is connected to the suction pipe and the delivery pipe. Power from external source is transferred to the piston through crank rod, connecting rod, etc., as shown in Fig. 13.2. A suction valve is provided between cylinder and suction pipe. Also delivery valve is provided between cylinder and delivery pipe.

Demonstration

(i) The rotation of the crank causes the plunger to move backward and forward in the cylinder.

(ii) Vacuum is created in the cylinder as the plunger moves to the right during suction stroke.

(iii) The atmospheric pressure on the water surface forces the water up the suction pipe. The suction valve is opened and the water enters the cylinder.

(iv) On the return stroke of the plunger (plunger moving to the left). called delivery stroke, the water pressure closes the suction valve and opens the delivery valve, forcing the water up the delivery pipe.

(v) Again in the suction stroke the water is sucked from the sump and the cycle continues.

Observations

Stroke length =

Stroke diameter =

Suction head =

Delivery head =

Speed =

Discharge =

Power required =

(c) Centrifugal Pump

Theory: Centrifugal pumps, also called dynamic pressure pumps, are those in which pumping of liquids is affected by rotary motion of one or more rotating wheels called impellers. The pump must be full while starting and should not be allowed to drain. The pump is driven by power from an external source, by which means the vanes are rotated. This gives a centrifugal head to the water in the pump and the water will leave the vanes at the outer circumference with high velocity and pressure.

A partial vacuum is formed in the centre, into which water from the suction pipe flows. High velocity of leaving water is transformed into pressure head by causing the leaving water to flow through a passage of gradually increasing area known as diffuser. The kinetic energy of the leaving water is thus converted into pressure energy, which is utilised in increasing the delivery head of the pump.

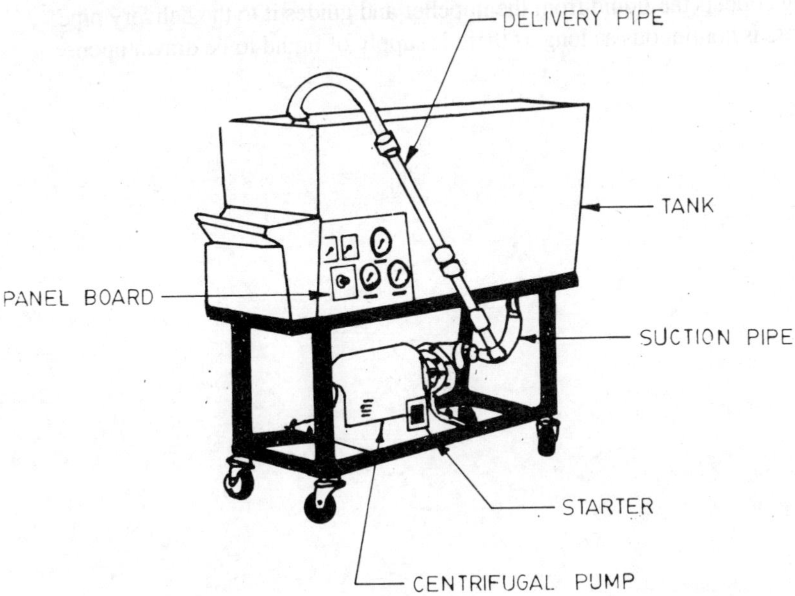

Fig. 13.3 Centrifugal Pump Demonstration Set

Apparatus: A working model of centrifugal pump is shown in Fig. 13.3. It consists of the following elements:

(i) Rotating element consisting of shaft and a vaned rotor called impeller. The vanes are curved, mostly cylindrical. The impeller is mounted on a shaft coupled to the driving unit which may be electrical motor or internal combustion engine.

(ii) Stationary element consisting of casing, stuffing box and bearings. The casing is an airtight chamber surrounding the pump impeller.

(iii) Suction pipe consists of strainer and foot valve. It connects the centre of the impeller to the sump from which the liquid is to be lifted and is provided with a strainer at its lower end so as to prevent the entry of solid particles into the pump. A foot valve is one-way valve located above the strainer into the suction pipe. It prevents backflow when the pump stops working.

(iv) Delivery pipe and delivery valve: It leads the liquid from the pump outlet to point of use. A regulating valve provided just near the pump outlet serves to control the flow of liquid into the delivery pipe.

Demonstration

(i) Prime the pump so that the suction pipe, casing and portion of the delivery pipe up to delivery valve are completely filled with the liquid to be pumped.

(ii) Switch on the pump so that motion imparted to the impeller builds up centrifugal force which throws the liquid towards the impeller periphery. Consequently, liquid from the pump is sucked in towards the impeller eye.

(iii) The casing collects the liquid from the impeller and guides it to the delivery pipe.

(iv) This process is continuous as long as there is supply of liquid to be drawn upon.

Observation

Rated head =

Discharge =

Normal speed =

Power required =

Size =

Type =

Impeller diameter =

EXPERIMENT 14

To demonstrate the actual working of the following:

(a) Pelton wheel turbine

(b) Francis turbine

(a) Pelton Wheel Turbine

Theory: Pelton wheel is an axial-flow impulse turbine in which the entire available head of water is converted into velocity by making it pass through a nozzle or nozzles kept close to the runner. The jet of water from the nozzle is made to impinge on the vanes. It enters the vanes at high velocity and, after flowing over them, leaves at low velocity thus imparting energy to the impeller. The pressure of water, both at entrance and exit of the vanes is atmospheric. Water is admitted over a part or the whole of the circumference of the impeller.

If h = height of the head-race from the foot of fall and

v = velocity of the jet of water from the nozzle

then $h = \frac{v^2}{2g}$, assuming that entire head is converted into velocity head.

$$\text{Energy absorbed by the runner per kg of water} = h - \frac{v_1^2}{2g} = \frac{v^2}{2g} - \frac{v_1^2}{2g}$$

where v_1 is the velocity of water at the exit from the vanes.

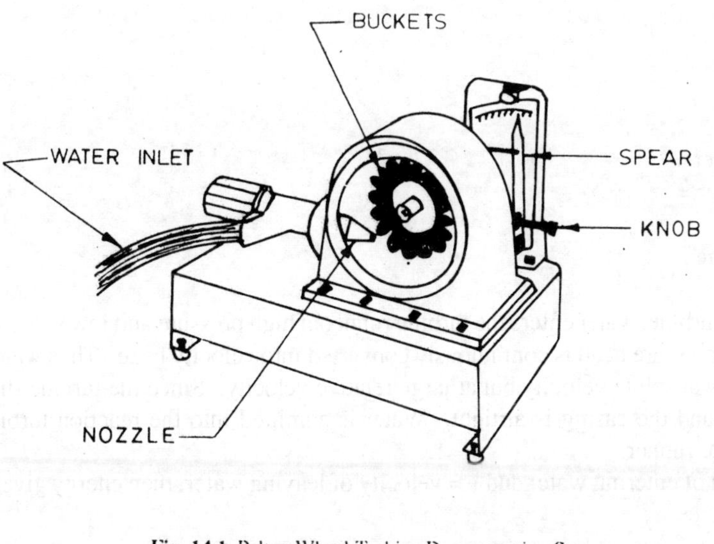

Fig. 14.1 Pelton Wheel Turbine Demonstration Set

Apparatus: Pelton turbine demonstration model consists of a pelton wheel turbine, a supply pump unit, flow measuring unit and piping system. The unit essentially consists of a casing with a large circular transparent window kept at the front for visual inspection of the impact of the jet on buckets as shown in Fig. 14.1. A rope-brake arrangement is provided to load the turbine. The input to the turbine can be controlled by adjusting the spear position by means of a handwheel fitted with indicator arrangement. Water inlet pressure is measured by a pressure gauge and a hand technometer is provided for measurement of speed.

Demonstration

(i) Make sure before starting that the pipelines are free from foreign matter and that all the joints are watertight and leakproof. The spear in the pelton wheel turbine inlet should also be in closed position.

(ii) Prime the pump and start it with closed gate valve.

(iii) Slowly open the gate valve situated above the turbine and open the cock fitted to the pressure gauge and see that the pump develops the rated head.

(iv) Now open the turbine spear by rotating the handwheel until the turbine attains the normal rated speed.

(v) Run the turbine at normal speed for about one hour. Load the turbine slowly.

(vi) Open the water inlet valve and allow some cooling water through the break-drum when the turbine runs under load, so that heat generated by the brake-drum is carried away by the cooling water.

(vii) Note down the technical details/specification of Pelton wheel turbine.

Observations

Rated supply head =
Discharge =
Normal speed =
Power output =
Unit speed =
Jet diameter =
Pitch circle diameter =
Jet ratio =
No. of buckets =
Brake-drum diameter =
Rope-brake diameter =

(b) Francis Turbine

Theory: In Francis turbine, water enters the turbine runner at high pressure and low velocity. As water flows over the vanes, the pressure head is continuously converted into velocity head. Thus water leaves the vanes at low pressure, low absolute velocity but at large relative velocity. Since the turbine runs under pressure, its wheels run full and the casing is airtight. Water is admitted into the reaction turbine over the entire circumference of the runner.

If h = total head of entering water and v = velocity of leaving water, then energy given to the runner per kilogram of water $= h - \dfrac{v^2}{2g}$

Apparatus: Francis turbine demonstration model consists of Francis turbine, a supply pump unit, flow measuring unit, and a piping system. A straight conical draft tube as shown in Fig. 14.2, is provided for the purpose of regaining the kinetic energy from the exit water and also facilitating easy accessibility of the turbine due to its location at a level higher than the tail race. A transparent hollow perspex cylinder is provided in between the draught bend and a casing for the purpose of observation of flow at exit of runner. A rope-brake arrangement is provided to load the turbine. The output of the turbine can be controlled by adjusting the guide vanes for which a handwheel and a suitable link mechanism is provided. The net supply head is measured by a pressure/vacuum gauge, and for measurement of speed a hand technometer is provided.

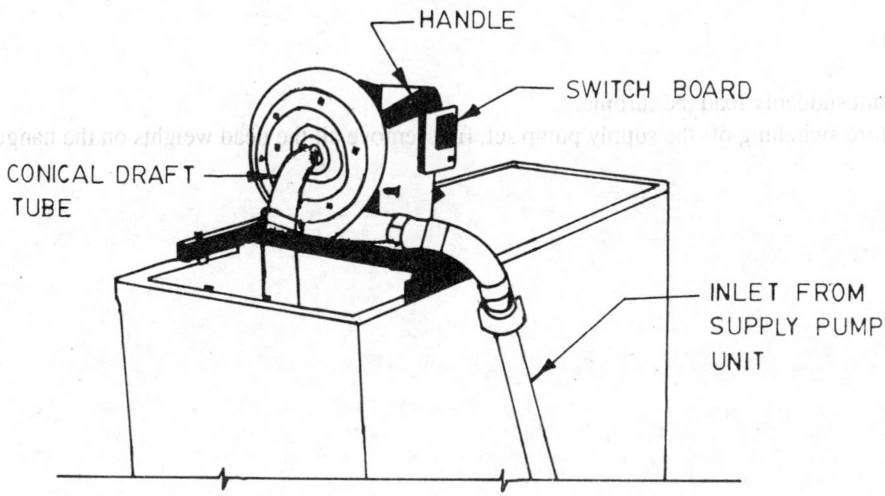

Fig. 14.2 Francis Turbine Demonstration Set

Demonstration

(i) Make sure before starting that the pipelines are free from foreign matter and that all the joints are watertight and leakproof. Guide vanes in the Francis turbine should also be in closed position.

(ii) Prime the pump and start it with closed gate valve.

(iii) Slowly open the gate valve situated above the turbine and open the cock fitted to the pressure gauge and see that the pump develops the rated head.

(iv) Now open the guide vanes by rotating the handwheel until the turbine attains the normal rated speed.

(v) Run the turbine at normal speed for about one hour. Load the turbine slowly.

(vi) Open the water inlet valve and allow some cooling water through the brake-drum when the turbine runs under load, so that heat generated by the brake-drum is carried away by the cooling water.

(vii) Note down the technical details/specification of Francis turbine.

Observations

Rated supply head =
Discharge =
Normal speed =

Power output =
Unit speed =
Specific speed =
Runaway speed =
Runner diameter =
No. of guide vanes =
Pitch circle diameter of guide vanes =
Brake-drum diameter =
Brake-rope diameter =

Precautions

(i) Do not suddenly load the turbine.
(ii) Before switching off the supply pump set, first remove all the dead weights on the hanger.

REFERENCES

The author has found the following books and Indian standards quite useful in his teaching of the subject of hydraulics, and in compilation of this laboratory manual.

- Massey, B S : *Mechanics of Fluids*, Van Nostrand.
- Streeter, V L : *Fluid Mechanics*, McGraw-Hill and Wylie E B.
- Shames, I H : *Mechanics of Fluids*, McGraw-Hill.
- Kumar, D S : *Fluid Mechanics and Fluid Power Engg.* Katson Publishing House.
- Francis, J R D : *Fluid Mechanics for Engg. Students*, Edward Arnold.
- Simon, A L : *Practical Hydraulics*, John Wiley.
- Govinda Rao, N S : *Mechanics of Fluids*, Orient Longman.
- Dake, J M K : *Essentials of Engg. Hydraulics*, The Macmillan Press.
- Metcalf and Eddy : *Waste Water Engineering*, McGraw-Hill.
- Fair Geyer Okun : *Water Supply and Waste Water Removal*, John Wiley.
- Hussain, S K : *Water Supply and Sanitary Engg.*, Oxford and IBH.
- IS : 3911-1966 : Specification for surface floats.
- IS : 4858-1968 : Specification for velocity rods.
- IS : 4477 (Part I)-1967 : Methods of Measurement of fluid flow by Means of Venturimeters.
- IS : 3910-1966 : Specifications for current meter (cup-type) for water flow measurements.
- IS : 3918-1966 : Code of practice for use of current meter (cup-type).
- IS : 1191-1971 : Glossary of terms and symbols used in connection with the measurement of liquid flow with a free surface.
- IS : 2951-1965 : Recommendation for estimation of flow of liquids in closed conduits.
- IS : 1194-1960 : Forms for recording measurement of flow of water in open channel.
- IS 2952 (Part I)-1964 : Recommendations for methods of measurement of fluid flow by means of orifice plates and nozzles.

APPENDIX - I

Table I (i) Properties of some common liquids at 20° C and atmospheric pressure

Fluid	Mass density (ρ) slug/m^3	Specific weight (w) kg/m^3	Bulk modulus (K) kg/cm^2	Surface tension (σ) kg/m	Vapour pressure (p_v) kg/m^2	Dynamic viscocity (μ) Poise	Dynamic viscocity (μ) $\frac{kgs}{m^2}$	Kinematic viscosity (ν) Stokes	Kinematic viscosity (ν) m^2/s
Water	101.95	999.00	2.205×10^{-4}	0.0075	0.024	1.00×10^{-2}	1.02×10^{-4}	1.00×10^{-2}	1.00×10^{-6}
Air	0.126	1.208	1.05	—	—	1.85×10^{-4}	1.88×10^{-6}	0.153	1.53×10^{-4}
Mercury	1380	13550.0	2.675×10^{5}	0.052	1.76×10^{-6}	0.016	1.63×10^{-4}	1.18×10^{-3}	1.18×10^{-7}
Glycerine	129.3	1260	4.44×10^{4}	0.0065	1.40×10^{-7}	8.35	8.5×10^{-2}	6.63×10^{-3}	6.63×10^{-7}
Carbon tetra-chloride	162.50	1594	1.126×10^{4}	0.0027	1.30×10^{-1}	0.01	1.02×10^{-4}	6.30×10^{-3}	6.40×10^{-7}
Kerosene	81.50	800	—	0.0024	—	0.02	2.04×10^{-4}	25.0×10^{-3}	25.0×10^{-7}
Benzene	89.80	880	1.056×10^{4}	0.0026	1.02×10^{-1}	0.007	7.15×10^{-5}	7.95×10^{-3}	7.95×10^{-7}
Castor oil	97.80	960	1.469×10^{4}	0.0040	—	9.80	1.0×10^{-1}	10.00	1.02×10^{-3}
Ethyl alcohol	80.4	789	1.23×10^{4}	0.0022	5.9×10^{-2}	0.012	1.22×10^{-4}	1.53×10^{-2}	1.53×10^{-6}

Table I (ii) Mechanical properties of water at atmospheric pressure

Temperature	Density $\frac{kg\text{-}s^2}{m^4}$	Specific weight kg/m^3	Dynamic viscocity poise	Surface tension g/cm	Vapour pressure kg/cm^2	Elastic modules kg/cm^2
0	101.9	1000	17.8×10^{-3}	.0771	.00623	2017
10	101.9	1000	13.1	.0758	.0125	2144
20	101.7	998	10.09	.0744	.0238	2234
40	101.1	992	6.54	.710	.0775	2330
60	100.2	983	4.78	.0676	.2038	2320
80	99.1	972	3.66	.0640	.475	2240
100	97.8	958	2.95×10^{-3}	.0601	1.035	2110

Table I (iii) Common liquids used in the Manometer

Manometer liquid	Relative density	Colour	Remarks
Water	1.0	Colourless	Corrodes iron and steel
Ethyl alcohol	0.794	Colourless	Corrodes iron and steel
Methyl alcohol	0.797	Colourless	Corrodes iron; poisonous; toxic;
Acetylene tetrabromide	2.964	Yellow (red for commercial use)	Corrodes iron and steel; weakens rubber; poisonous
Kerosene	0.800	Colourless	Noncorrosive; toxic; poisonous
Mercury	13.56	Silver	Combines with brass and copper; poisonous; toxic